Factoring the Gravitational Redshift of Active Galactic Nuclei

toward clarification of the most mysterious objects in the universe

Robert Gnaski Burton

Apeiron
Montreal

Published by C. Roy Keys Inc.
4405, rue St-Dominique
Montreal, Quebec H2W 2B2 Canada
http://redshift.vif.com

First Published 2018

Library and Archives Canada Cataloguing in Publication

Burton, Robert Gnaski, 1945-, author
Factoring the gravitational redshift of active galactic nuclei : toward clarification of the most mysterious objects in the universe / Robert Gnaski Burton.

Includes bibliographical references.
Issued in print and electronic formats.
ISBN 978-1-987980-15-8 (softcover).--ISBN 978-1-987980-16-5 (PDF)

1. Active galactic nuclei. I. Title.

QB858.3.B87 2018 **523.1'12** **C2018-906546-X**
C2018-906547-8

Cover graphic: ESO image 0903A of the Seyfert 2 galaxy Centaurus A in constellation Centaurus (modified)

Table of Contents

Preface

Everybody loves a good mystery story. Perhaps the most intriguing mystery concerns the creation of the Universe. The story has been told many times, in many different ways, with both religious and scientific appeal. Each episode offers a tantalizing new discovery, but the mystery always remains with us. Nobody knows for sure how it will end. This book gives a different version of the episode about active galactic nuclei, the most mysterious objects in the Universe. The new version disputes their measured luminosities, and it establishes a more consistent evolutionary model based upon recent observations and spectrographic evidence. The technicalities are aimed at advanced undergraduate students in astronomy and astrophysics, but the text is sufficiently intermixed with material easily understood by the layperson. Although many fundamental questions will remain unanswered, all readers are given to understand that we have been looking at these distant(?) objects through a distorted lens.

Given my lack of expertise in the wide range of subject matter, it was inevitable that I would depend heavily on many journals and research articles to develop my material. This is evident by the hundreds of citations found within the text. I trust that I will not be too severely criticized for the exclusion of many other works which are equally worthy of mention. Included are more than one hundred figures and tables from various publications that should help the reader better understand everything. Appendix A lists all the appropriate credits and permissions. I thank all the individuals who kindly provided their permissions for reuse of their copyrighted material. I thank in particular all those who took the time to answer questions or review a few paragraphs to ensure the accuracy of my work. I owe a particular debt of thanks to Martin Lopez-Correidora, Benny Trakhtenbrot, Yue Shen, Andreas Schulze, Hagai Netzer, Gabor Worseck, Xiaohui Fan, Martin Gaskell, Daniel Proga, Richard Plotkin, Gordon Richards, Jayant Narlikar, and Jack Sulentic. My warmest thanks go to those who provided technical editing for one or more of the chapters. Thank you ever so much, Chris Fulton, Dan Mayes, and Paola Marziani. I am grateful that I was able to find a gifted publisher, Roy Keys, who could help hone my prose in a professional manner. I am eternally grateful to my advisor, Dr. Melodi Rodrigue, who emboldened me to persevere for so long in what seemed an impossible task.

Robert Gnaski Burton
Reno, Nevada
October 23rd 2018

Acronyms and Abbreviations

The following acronyms and abbreviations are found in the text:

1PM	first post-Minkowskian order
2dFGRS	2dF galaxy redshift survey
2MASS	two micron all sky survey
2MRS	2MASS redshift survey
2QZ	2dF QSO redshift survey
3DE1	three dimensional E1 (parameter space)
4DE1	four dimensional E1 (parameter space)
ADAF	advection dominated accretion flow
ADIOS	advection dominated inflow-outflow solution
AGN	active galactic nucleus
ApJ	astrophysical journal
ALMA	Atacama large millimeter array
BAL	broad absorption line (quasar)
BC	broadline component
BHMF	black hole mass function
BLR	broad-line region
CASTLES	CfA-Arizona space telescope lens survey
CDAF	convection dominated accretion flow
CDF	Chandra deep field
CDM	cold dark matter
CH	cosmological hypothesis (of AGN redshift)
CMB	cosmic microwave background
COS	cosmic origins spectrograph
COSMOS	cosmic evolution survey
DIR	decreasing intrinsic redshift (model)
DLA	damped Lyman-alpha (system)
DM	dark matter
DR3	SDSS data release 3
E1	eigenvector 1
EELT	European extremely large telescope
EGADS	extreme gravity from aggregated dark subhalos
eV	electron-volt
EW	equivalent width
FIT	Fackerell, Ipser and Thorne
FLRW	Freedman-Lemaitre-Robertson-Walker cosmology
FOS	faint object spectrograph
FPBH	fundamental plane of black holes

FR	Fanaroff & Riley (morphology)
FUSE	far ultraviolet spectroscopic explorer
FWHM	full width half maximum
G	gauss
GALEX	galaxy evolution explorer
GTC	*Gran Telescopio Canarias*
GMT	giant Magellan telescope
GUT	grand unification theory
Gyr	gigayear
HE	Hamburg ESO
HF	Hoyle & Fowler
H_I	neutral hydrogen
He_I	neutral helium
He_{II}	singly-ionized helium
He_{III}	doubly-ionized helium
HERS	helium reionization survey
HiBAL	high-ionization BAL
HIL	high ionization line
HST	Hubble space telescope
HZT	high-z supernova search team
ICM	intracluster medium
IGM	intergalactic medium
ILR	intermediate-line region
IR	infrared
ISAAC	infrared spectrometer and array camera
ISCO	innermost stable circular orbit
JWST	James Webb space telescope
keV	kiloelectron-volt
L_{bol}	bolometric luminosity
L/L_{Edd}	relative accretion rate
LAF	*Lyman-alpha forest*
ΛCDM	*lambda* cold dark matter
LF	luminosity function
LH	local hypothesis (of AGN redshift)
LIL	low ionization line
LoBAL	low-ionization BAL
MACHO	massive compact halo object
M_{BH}	black hole mass
MC	magnetic coupling
MHD	magnetohydrodynamics
MNRAS	Monthly Notices of the Royal Astronomical Society
Mpc	megaparsec

Myr	megayear
NIR	near infrared
NLR	narrow-line region
pc	parsec
PC	principal component
PCA	principal component analysis
QSO	quasi-stellar object (quasar)
QSSC	quasi-steady state cosmology
QTC	quantum temporal cosmology
R_H	Holmberg radius
RIAF	radiatively inefficient accretion flow
RL	radio loud
RLF	radio loud function
RLQ	radio loud quasar
RM	reverberation mapping
RM	rotation measure
RMS	root mean square
RQ	radio quiet
RQQ	radio quiet quasar
RRM	residual rotation measure
R_S	Schwarzchild radius
SBF	surface brightness fluctuation
SCP	supernova cosmology project
SDSS	Sloan digital sky survey
SED	spectral energy distribution
SF	star formation
SMBH	supermassive black hole
SNe Ia	Type-1a supernova
TF	Tully-Fisher relation
TMT	thirty meter telescope
ULIRG	ultraluminous IR galaxy
VBLR	very broad line region
VH	variable hypothesis (of AGN redshift)
VLBA	very long baseline array
VLBI	very long baseline interferometry
VLT	very large telescope
VRRM	velocity-resolved reverberation mapping
WIMP	weakly interacting massive particle
WPK	Welter, Perry & Kronberg
XRB	X-ray background

Chapter 1

Measurement of Distant Galaxies

The majesty of the night time sky has fascinated humankind since the dawn of civilization. The grandeur of the heavenly stars inspires contemplation upon the great metaphysical questions. How large is the universe? Did it have a beginning? What is my place within its vast reach? The ancients posed answers deeply rooted in mythology. Modern civilization leans more heavily upon science for solutions. The science of astronomy has developed the means to look deep into space at trillions of extragalactic objects and calculate their mass, luminosity and chemical composition. It also has the means to calculate their distance from Earth. We now can identify sources of light that have taken billions of years to reach our eyes. Such measurements are crucial in answering the questions of the size and (possible) origin of the cosmos.

When he used the first astronomical telescope in 1609, Galileo Galilei had no means to determine the distance to any of the millions of new stars he discovered lying in the plane of the Galaxy (known today as the Milky Way). It was over two centuries until Friedrich Bessel made the first successful distance measurement using a method called *parallax*. As the Earth revolves around the Sun, a nearby star traces out an ellipse in its apparent motion against any background star. Given the size of the semi-major axis of the ellipse and the known diameter of Earth's orbit around the Sun, it is possible to calculate trigonometrically the distance to the nearby star. The closest star to the Sun is at a distance of approximately four light-years. This subtends a parallax angle less than one arcsecond, which translates to a distance of 3.26 light-years. The available angular resolution of 5 milliarcseconds allows distance measurements of no greater than 200 parsecs or 650 light-years. Galileo would have been surprised at the distances he was observing in his telescope. He probably would have been more surprised to learn that a distance of roughly one parsec separates any two given stars in the 'thick carpet' of the Milky Way.

Robert Gnaski Burton, (Montreal: Apeiron 2018)

Distances to only 1% of the one hundred billion stars in the Galaxy can be measured by means of parallax. Another method would be necessary to measure the remaining 99%. In 1912 Henrietta Leavitt discovered that Cepheid variable stars could be used as *standard candles*. The Cepheid standard candle is a variable star whose pulsational period has a strong correlation to its intrinsic luminosity. Since light from a point source drops off according to the inverse square law, a simple calculation reveals the distance to the source. Use of the Cepheid variable provides ground-based measurements to a distance of 4 Mpc (four million parsecs), within the range of the entire Galaxy. Unknown to most astronomers at this time was that there was a much larger world out there yet to explore.

Early twentieth century astronomy is noteworthy for the vigorous debate over the *spiral nebulae*. Some astronomers argued that these faint misty patches of light were giant groups of stars within the Galaxy, while others believed they were 'island universes' of stars at extragalactic distances. The resolution came in 1924 from Edwin Hubble's observations using the newly constructed 100-inch reflecting telescope at Mount Wilson Observatory. His observations showed conclusively that the spiral nebulae were islands of stars located very distant from our Galaxy. Furthermore, they were isotropic on the sky and *all receding away* at a rate proportional to their distance from Earth. Hubble summarized this astounding discovery in his 1936 book in which he wrote:

> ...[But] nebular spectra are peculiar in that the lines are not in the usual positions found in nearby light sources. They are displaced toward the red of their normal position, as indicated by suitable comparison spectra. The displacements, called red-shifts, increase, on the average, with the apparent faintness of the nebula that is being observed. Since apparent faintness measures distance, it follows that red-shifts increase with distance. Detailed investigation shows that the relation is linear.[1]

Today this apparent linear relationship is known as *Hubble's Law*:

$$v = H_0 D,$$

where H_0 is the constant of proportionality between the apparent radial velocity v (calculated from the spectral redshift, z) and the distance D. In theory the distance to any extragalactic object becomes measurable on condition of an accurate spectrograph and the value of H_0. With this great discovery extragalactic astronomy was born. In less than one hundred years the worldview of mankind had changed beyond all imagination. No longer did we live under a panoply of stars several light-years distant. Our world was vastly larger and getting larger all the time. Many profound conclusions about large-scale structures and the history of the universe hinge on the interpretation of 'The Law'. It is therefore crucial to understand its relative degree of accuracy and the conditions under which it remains valid. Three important questions need precise answers.

The First Question: How Accurate is H_0?

Establishing an exact value for H_0 is an elusive goal. For decades computed values fell between 50 and 100 km s^{-1} Mpc^{-1}, giving rise to the 'factor-of-two' controversy. Many redshift and distance measurements are necessary in order to converge on its value. Although the acquired galactic redshifts are reasonably accurate, the measured galactic distances are often spurious. Astronomical distances are determined in ladder fashion. Methods appropriate over short distances are used to calibrate techniques used for deeper space. Parallax is used to calibrate standard candles (Cepheids, supernovae Ia, *etc.*), and standard candles are used to help calibrate extragalactic techniques. Each step along the way is subject to error. This makes the final computation of H_0 subject to error from any and all steps along the distance ladder. By the end of the 20th century, standard candles still could not derive distances to Local Group galaxies more accurately than 10%.[2] While Cepheid variables remain a powerful tool for measuring the distance to nearby galaxies, other methods must be applied for deeper space. The most notable of the methods for distant galaxies are the Tully-Fisher (TF) relation for spiral galaxies, the surface brightness fluctuation (SBF), and the maximum luminosities of Type Ia supernovae (SNe Ia). Each method introduces some level of systematic error, rendering calibration of H_0 imprecise. Freedman and Madore (2010) applied all of these methods to get a converged estimate of $H_0 = 73 \pm 4$ km s^{-1} Mpc^{-1}.[3] This represents a 5% systematic uncertainty in the Hubble constant. Further constraint is necessary for more precise measurements of cosmological quantities.

The Second Question:
Is The Hubble Law Universal in Nature?

Two important factors disrupt the direct correspondence between the recessional velocity and distance. First, the recessional velocities of nearby galaxies are indistinguishable from the random velocities induced by gravitational interactions with their neighbors. Therefore, the Hubble Law is not applicable for distances much less than 10 Mpc ($z < 0.002$).* Second, any deceleration/acceleration of cosmic expansion will cause a non-linear decrease/increase in the recessional velocity. Many astronomers in the late twentieth century expected a slowdown in the expansion rate due to the universal attractive force of gravity. It would not be possible to modify the Hubble Law until there was an accurate way to measure universal expansion.

The modern theoretical relationship between recessional velocity and redshift ultimately depends upon the choice of cosmological model. Most popular is the Friedmann-Lemaitre-Robertson-Walker (FLRW) cosmology used to describe the expansion of a *homogeneous* and *isotropic* universe. In FLRW the luminosity distance (in Mpc) is given by:†

$$D_L = cH_0^{-1}(1+z)\left|\Omega_k\right|^{-\frac{1}{2}} sinn$$
$$\left\{\left|\Omega_k\right|^{\frac{1}{2}} \int_0^z dz\left[(1+z)^2(1+\Omega_M z) - z(2+z)\Omega_\Lambda\right]^{-\frac{1}{2}}\right\},$$

where Ω_M is the mass density (including dark matter), Ω_Λ is the vacuum energy density (dark energy), $\Omega_k = 1 - \Omega_M - \Omega_\Lambda$ (implying a flat universe), and *sinn* is *sinh* for $\Omega_k > 0$ and *sin* for $\Omega_k < 0$. The simple linear law of Hubble has become a very difficult FLRW equation requiring numerical analysis for solutions.

Luminosity distance is given exactly by $d_L = (L/4\pi F)^{1/2}$, where L is the intrinsic luminosity and F is the observed flux (integrated over all frequencies). When accurately observed at large distance, d_L can be substituted for D_L in the FLRW equation, and a family of solutions will ensue for the combined cosmological parameters of Ω_M and Ω_Λ. By the late 1990s it became possible to obtain accurate luminosity measurements of Type Ia

* When recessional velocity is much less than the speed of light, $v \ll c$, then $z = v/c$.

† FLRW derivations rest upon the equations of General Relativity, which is outside the scope of this book.

supernovae at distances reaching $z \sim 0.5$. This set the stage for projects to calculate Ω_M and Ω_Λ and also to refine Hubble's Law.

Two independent research groups went on the trail of distant Type Ia supernovae. First to report was the High-Z supernova search Team (HZT) with observations of ten Type Ia's in the redshift range $0.16 < z < 0.62$.[4] Soon afterward the Supernova Cosmology Project (SCP) reported on the analysis of forty-two Type Ia's in the range $0.1 < z < 0.83$.[5] Most remarkable in their reports was their high confidence level that Ω_Λ had a significant non-zero value. In other words there must be some mysterious force (dark energy) causing the expansion velocity to *increase*. Bizarre as it first seemed, further support came from the independent observation of the angular power spectrum of the cosmic microwave background (CMB) fluctuations. The worldview once again underwent a dramatic change. Science history author Helge Kragh writes:

> …the universe of modern cosmology hardly resembles the one we see. It is composed to a first approximation of invisible dark energy whose physical origin remains obscure. Most of what remains is in the form of cold dark matter particles, whose identity is also unknown. Close inspection is needed to make out the further contribution of neutrinos, although this too is nonzero. And baryons, the stuff of which we are made, are little more than a cosmic afterthought. This picture entails a shift in the way we see the universe that is profound enough to be called a Copernican counter-revolution.[6]

Recognition of the dark energy phenomenon implies that the Hubble Law cannot be universally applicable. There comes a distance at which the linear relationship to redshift no longer holds. That distance is not known for certain, but reasonable estimates are possible with the use of the FLRW luminosity distance equation given above. Given H_0 and the assumption of a flat universe, a plot of distance *vs.* redshift shows the curvature for any theoretical value of the mass density, Ω_M.[7] (See Figure 1.1.)

Before the *discovery* of dark energy, luminosity distance was expected to decelerate for increasing redshift. Instead, the (unforeseen) vacuum energy overcomes the attractive force of gravity leading to the positive curvature. The smaller the mass density, Ω_M, the greater the vacuum energy, Ω_Λ, and the greater the acceleration.

It is important to take note of the added degree of uncertainty in the revised distance calculation. Hubble's Law resolves distances at hundreds

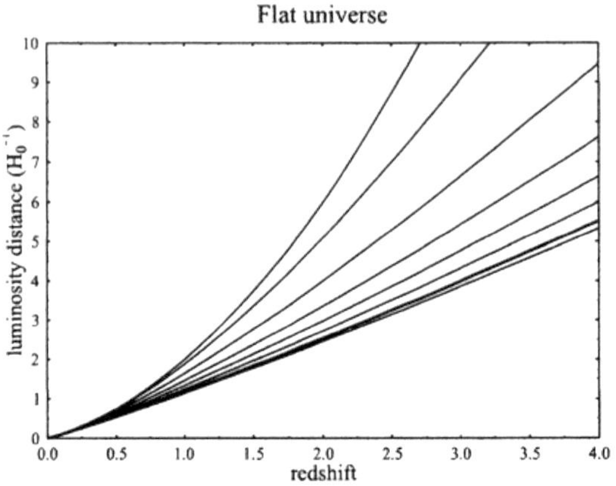

Figure 1.1 - FLRW luminosity distance (D_L) vs. redshift (z)
Ω_M = 0, 0.05, 0.2, 0.4, 0.6, 0.8, 1.0, 1.05 (top to bottom)

of megaparsecs ($0.001 < z < 0.2$) where the uncertainty rests upon systematic errors inherent in the observation of standard candles. The FLRW equation provides a measurement to the ends of the universe, but it rests further upon the assumptions of the FLRW cosmology, limited measurements of Type Ia supernovae (with its significant spread in peak luminosity)*, and undetermined values for Ω_M and Ω_Λ. Its level of uncertainty is underscored by the fact that technical papers always call out their preferred values for H_0, Ω_M and Ω_Λ in any discussion pertaining to high-z distance measurement.

The Third Question: Do Some Extragalactic Objects Exhibit Intrinsic Redshift?

The cosmological hypothesis states that the universe is isotropic and homogeneous, and that all extragalactic objects derive their redshifts from the expansion of the universe. This is a most difficult principle to prove in fact. The universe is a very large place, perhaps infinite in nature. It simply is not possible to prove that every radiating object is devoid of some redshift internal† to its nature. If some intrinsic feature causes 2-3% of the redshift, the uncertainty in the Hubble constant would prevent its detection. A yet larger fraction of intrinsic redshift would go undetected for high-z objects due to the uncertainty in the FLRW parameters. A more practical

* See (Bothun 1998), [2] pg. 61.

† 'internal' is meant to exclude 'cosmological' and 'Doppler' (redshift).

means to answer the third question is to attempt a disproof of the cosmological redshift hypothesis. This would be achieved by proving that some class of objects contains intrinsic redshift that is outside the margin of error in current astronomical measurements. Since redshift in the FLRW equation is based solely on cosmic expansion, any additional redshift would overvalue the measured intrinsic luminosity. Objects would seem brighter in the sky than they actually are. If such 'illusory' brightness does exist in the observable sky, it most likely comes from the class of objects known as quasars (quasi-stellar objects, or QSOs).*

The first identified quasar was 3C 273 in the constellation Virgo. 3C 273 is a source with two components (see Figure 1.2) separated by 20 arcseconds. In 1962 its position was obtained with an accuracy of ~1 arcsec. Its spectrum presented quite a puzzle, for it did not resemble anything with typical stellar chemical elements. In the following year Maarten Schmidt solved the mystery when he noticed that the spectral lines of hydrogen were redshifted to a calculated value of $z = 0.158$.† Astronomers had detected an object billions of light-years from Earth!

If 3C 273 is at its cosmological distance, its mass is nearly 1 billion solar masses ($M_{3C273} \sim 10^9\ M_{Sun}$), and its luminosity is greater than 4 trillion times that of the Sun ($L_{3C273} > 4 \times 10^{12}\ L_{Sun}$), hundreds of times brighter than the entire Milky Way. Also, the diameter of 3C 273 is less than one light-year, while the Milky Way is 100,000 light-years across.

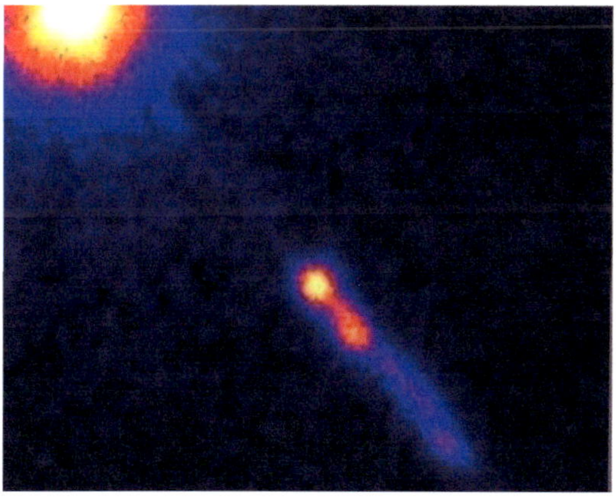

Figure 1.2 – Quasar 3C 273

These bizarre values presented a dilemma to the best minds in astronomy. Any dense conglomeration of stars does not explain such enormous energetic output. The first idea was some sort of massive central object with ac-

* Quasar (or Quasi-Stellar Object, QSO) is a term originally applied only to the starlike counterpart of a radio source whose redshift is much larger than that of normal galaxies. The term now applies regardless of any radio emissivity.

† The currently accepted redshift value of 3C 273 is $z = 0.173$.

cretion-powered radiation. The early theories of Hoyle, Fowler, Salpeter, Zel-Dovich and Lynden-Bell developed this idea into the current model of accretion onto a supermassive black hole (SMBH) at the dynamical center of its host galaxy. Further research has confirmed this model and showed that the galaxies are similar to those discovered and cataloged in the 1940's by astronomer Carl Seyfert. Quasars and 'Seyferts' both have nuclear emission lines superposed on a normal star-like spectrum that are very much broader than the widths of the absorption lines. They also have continuity in other properties including large UV flux, X-ray sources, and time-variable continuum flux. Seyferts and quasars are now considered part of a broader unified model of AGN that comprises many disparate subclasses of active galaxies. An *active galaxy* is one in which a significant fraction of its total luminosity is radiation not ultimately attributed to stellar photospheres. Researchers are inconsistent with the dividing line between Seyferts and quasars. Some delineate on the basis of luminosity ($L_{bol} = 10^{45}$ erg s^{-1}); others, the basis of redshift. This book uses the latter approach. A Seyfert is an instance of a closer AGN whose redshift is less than 3C 273; a quasar is an instance of a more distant AGN whose redshift is greater than that of 3C 273. Given that Seyferts and quasars are different manifestations of the same phenomena, the properties of the less distant Seyferts can be studied to help unlock the secrets of the mysterious quasars.

Active galactic nuclei have become quite the hot area of research. More than 1000 articles on the subject are published annually in refereed journals. There is a lot out there to explore. There are now over one million known AGN sources, and ~2% of galaxies at $z < 0.1$ have a fast-accreting supermassive black hole. Their extraordinary luminosity* and distance make these objects a very big story indeed. The scientific explanation is still not well understood, and their presence in the early universe flouts our best cosmological models. Of course the big story may yet be that they are not as far away as they appear. They may contain some intrinsic redshift factor that distorts our distance calculations. The worldview certainly would alter if any intrinsic component were identified. The aim of this book is to examine the causes of AGN redshift and to explore possible intrinsic factors.

* $L_{Sun} \sim 3.9 \times 10^{26}$ watts, whereas $L_{quasar} \sim 9.75 \times 10^{36}$ to 1.95×10^{39} watts.

Most astronomers today are convinced that quasars are at their cosmological distances. There are good arguments for their belief, and these are covered in Chapter 2. Other astronomers remain doubtful. Chapter 3 discusses the difficulties with the cosmological redshift hypothesis and offers counterarguments to it. Some astronomers have shut the door on the possibility of any galactic intrinsic redshift. That door should remain ajar if a *plausible* theory for it exists. Chapter 4 presents a theory based on existing physical principles that predicts variable, at times substantial, intrinsic redshift for quasars. Recent observational evidence in support of this new theory is the subject of Chapter 5. Finally, its plausibility is evaluated in Chapter 6 by comparison with statistically significant distance-independent data.

Chapter 2

The Case in Favor of the Cosmological Hypothesis

The discovery of the *anomalous* quasars in the 1960s led to much controversy in the community of astrophysicists. Many theories were advanced to explain the large redshift measurements. Some sustained the existing redshift/distance relationship of the cosmological hypothesis (CH). Others countered with a *local* hypothesis (LH) for the quasars on the basis that no existing framework could explain their superluminosity at such great distances. Many conferences were held over the next two decades, during which the arguments were intense and personal. In 1983 the National Research Council charged the Astronomy Survey Committee with making recommendations concerning the facilities needed for progress in astronomical research. It declared that "the conventional interpretation of redshift, as velocity related and indicative of distance in an expanding universe, provides the most natural and self-consistent explanation within the framework of the Big Bang Theory." [8] By way of this recommendation, leading LH proponents were denied further viewing time at the Palomar Observatory. Such a serious action led the astronomical community at large to view the redshift debate as closed.

The majority opinion had been made, but the debate did not end. Objections to CH are found in journals all the way to the present day. In particular, Fred Hoyle, Geoffrey Burbidge and Jayant Narlikar* presented the case for a *variable* hypothesis (VH). [9] In their view too many close galaxy/quasar associations are seen in the sky. In some cases it must be the case that there is a variable intrinsic redshift z_i such that $(1 + z_Q) = (1 + z_G)(1 + z_i)$, where z_Q and z_G are the redshifts of the quasar and galaxy respectively.†

* Fred Hoyle and Geoffrey Burbidge (along with William Fowler and Margaret Burbidge) are known for their famous work that showed elements heavier than helium are created (synthesized) in the stars.

† See Chapter 4 for further details.

The objections to the cosmological redshift hypothesis "only died when proponents died or at least retired." [10] Nevertheless, most scientists strongly adhere to CH. They base their conviction on the interpretation of three main types of observations: (1) quasars are measured at the same distance as nearby neighbor galaxies; (2) quasars are gravitationally lensed behind distant galaxies; (3) Lyman-alpha absorption indicates intervening clouds along the path to the distant quasars.

Near Neighbor

It would seem that the most obvious method to support the cosmological hypothesis for AGNs would be to measure the redshifts of actual galactic neighbors. Indeed, this may be the only method available for estimating absolute brightness of quasars (QSOs)* if they were to possess any intrinsic redshift. For CH the redshift of the neighbor, z_G, should be close in value to the redshift of the QSO, z_Q. If we find most measurements yield $z_G = z_Q$, we will favor the cosmological hypothesis. If we find most yield $z_G \ll z_Q$, we will favor the local hypothesis. Alas, nothing in the field of astronomy is so simple. Whatever limits are chosen for the separation angle and whatever limits are chosen for the apparent magnitude, the survey will always yield some mix of redshift equalities and inequalities between the AGN and its *near neighbor*. Statistical analysis is necessary to form any definite conclusions.

Stockton (1978) pioneered the study of spectroscopy for QSO neighbors. He used the Burbidge, Crowne and Smith catalog to choose QSOs and galaxies for analysis of any potential redshift differential between them. The sample objects were restricted to declination $-15^\circ < \delta < +55^\circ$. QSOs were eliminated from the study if their redshifts were greater than those of the brightest sample normal galaxies or if the QSO magnitudes were not substantially brighter than those of the brightest sample normal galaxies at the same redshift. This yielded 27 QSOs with the values $z < 0.45$ and $m_v < 19.12 + 5\log(z)$. Galaxies were included if they met the magnitude limit of the Red Sky Survey prints and if they were within 45 arcsec of the chosen QSO. These criteria yielded 29 galaxies in 17 of the 27 QSO fields. Definite redshifts were obtained for 25 of the 29 galaxies. The QSO/galaxy

* The terms quasar and QSO will be used interchangeably in accordance with the referenced publications.

pair was considered *near* if their respective redshifts were within 1000 km s^{-1} of each other. Table 2.1 shows the revealing data from Stockton. [11]

			Galaxy Position			
QSO (1)	z_{QSO} (2)	Galaxy No. (3)	r(") (4)	θ(°) (5)	$z_{Galalxy}$ (6)	ΔV_r (km s^{-1}) (7)
0137+060	0.396	1	34	40	0.2303	–35200
1004+130	0.240	1	34	232	0.2423	500
		2	45	273	0.2400	0
1048–090	0.344	1	23	55		
		2	26	187	0.1255	–50900
1128+315	0.289	1	7	133	0.2896	100
		2	28	263		
		3	35	200	0.2920	700
1150+497	0.334	1	21	138	(0.2889)	(–9600)
1208+322	0.388	1	10	83	0.466	14500
		2	34	46	0.1494	–53500
1223+252	0.268	1	34	163	0.0911	–43800
1302–102	0.286	1	34	100	0.1458	–33300
		2	39	197	0.0940	–47000
1510–089	0.361	1	25	104	0.2536	–23000
1512+370	0.3708	1	11	107	0.3722	300
1525+227	0.253	1	33	188	0.2519	–300
		2	40	5		
1548+114	0.4358	1	9	260	0.4338	–400
		2	12	251	0.4327	–600
2135–147	0.2005	1	6	126	0.1997	–200
		2	16	332	0.2002	–100
		3	43	138	0.2008	100
2141+174	0.213	1	38	271	0.2106	–600
2251+113	0.323	1	28	259	0.3228	0
		2	41	296	0.3312	1700
2305+187	0.313	1	7	358	0.2427	–15600
		2	35	341	0.2424	–15600
2308+098	0.432	1	9	319	0.1726	–56000

Table 2.1 – Stockton QSO/galaxy pairs

The redshifts of the QSOs and galaxies are given in columns 2 and 6 respectively. Column 7 gives the calculated radial velocity difference, in the rest frame of the QSO, between the galaxy and the QSO in whose field it lies. Half of the galaxies have velocities within 1000 km s^{-1} of their respective QSOs. The probability that eight or more fields should show redshift agreements between QSOs and galaxies by chance is $< 1.5 \times 10^{-6}$. Stockton concludes, "A probability of this order constitutes virtual proof that QSO redshifts are cosmological in origin." Peterson (1997) singles out the significance of Stockton's work for demonstrating *definitively* that QSOs are at cosmological distances implied by their large redshifts. [12]

Left unmentioned in Stockton's paper are the data with ΔV_r greater than the absolute value of 1000 km s^{-1}. It is extremely unlikely to have so many distant galaxies with a line of sight onto the QSO, and nothing in the selection process explains the heavy negative skewing. What should we conclude? If the data do have statistical significance, then it is important to note that there are as many occurrences of $z_G < z_Q$ as there are $z_G = z_Q$.

Yee and Green (1984) offered a less direct statistical study in which fields of galaxies around 108 QSOs with redshifts $0.05 < z < 2.05$ are examined and compared with control fields 1° north of each QSO.[13] For $z < 0.45$ there are significantly more galaxies in the quasar fields than in the control fields. The surface density of galaxies shows a correlation with angular distances to the QSOs. The excess galaxy objects become fewer in number and decrease in apparent brightness with larger redshift. Altogether this implies object distances that are consistent with those of normal galaxies at cosmological distances of the quasars. Their paper does not contain any skewed data, but their analysis is indirect and limited to $z < 0.5$.

Researchers have offered other indirect proofs for CH. In their discussion of 17 quasars at $z < 0.5$, Gehren *et al.* (1984) showed the absolute magnitudes and diameters of the host galaxies are consistent with quasars at cosmological distances.[14] For 1 arcsec resolution optical imaging of 78 QSOs at $0 < z < 0.7$, Hutchings *et al.* (1984) showed a number of cases where nearby objects are almost certainly at the same distance as the QSO.[15] None of the supporting research shows distance relationships at $z > 0.5$, and nobody disproves the local hypothesis better than Stockton.

Many astrophysicists claim the 'definitive' work of Stockton (1978) as a pillar of support for CH. It is important to note, however, that ongoing research continues to provide statistically significant data that call his conclusions into question. For example, Burbidge and Napier (2009) used ~4000 QSOs from the Sloan Digital Sky Survey catalog and compared them with galaxies from the RC 3 galaxy catalog.[16] They show a large and systematic excess of close QSO-galaxy pairs (where $z_Q \gg z_G$).* Neither identification bias nor any sensible model of microlensing is capable of accounting for such an overabundance of the close pairs.†

---

* See Table 3.4 for a list of 40 close QSO-galaxy pairs.

† See Chapter 3, **Near Neighbor – Revisited**, for more counter arguments to Near Neighbor.

We are led to conclude:

1) Near Neighbor can disprove the local hypothesis (LH) of AGN redshift.
2) Near Neighbor cannot disprove the variable hypothesis (VH) of AGN redshift.

Gravitational Lensing

Another method to confirm the *minimum* distance to a galactic source object is observation by gravitational lensing. This can occur when the source and a foreground object (known as the lens) are on the same line of sight. Light from the distant point source passes very near the gravitationally distorted space of the massive lens object, whereby the light path can take several different trajectories to the observer. One or more of these paths may suffer from significant absorption. Thus, the most important quasar lens surveys to date have been conducted at radio wavelengths where such extinction is negligible. When the source and observers are sufficiently far from the lens mass, bending from both sides can produce double images or even a ring. Figure 2.1 illustrates the altered line of sight caused by a gravitational lens.

Distant sources are better potential candidates because the efficacy of a lens increases with the distance to the background object. Since the foreground galaxy does not image as an ideal point mass, the (quasar) source image is altered in angular size, shape and brightness. This can make its identification quite complex. Nevertheless, quasars in deep space provide a good set of source objects for gravitational lensing.

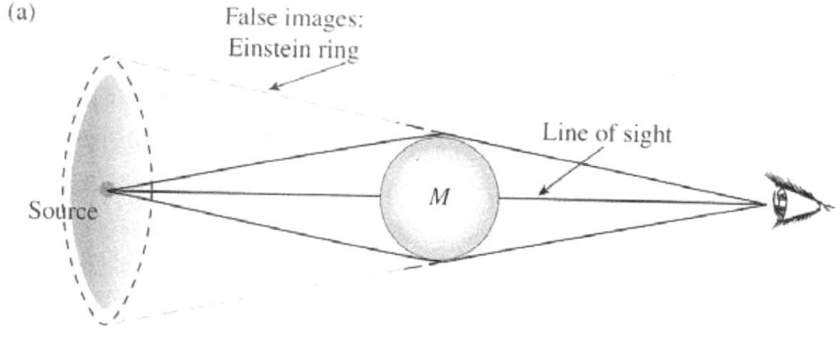

Figure 2.1

Figure 2.2 – Double-Quasar Q0957+561

Figure 2.3 – Quadruple J1000+0221

For 60 years the deflection of light by the Sun was the only known example of gravitational lensing. That changed when Walsh, Carswell and Weymann (1979) used the Kitt Peak National Observatory 2.1 meter telescope to discover a quasar multiply-imaged by a massive foreground cluster galaxy (see Figure 2.2). Gorenstein *et al.* (1984) provided evidence that the two sources Q0957+561 A and B, separated by 6 arcsec on the sky, are images of a single quasar whose radiation has been deflected by the gravitational field of a foreground cluster of galaxies.[17] Lens models that account for the A and B images further predict the existence of a third image with lower flux density. Thus, the QSO at $z_Q = 1.413$ appears more distant than the lensing galaxy at $z_G = 0.39$. This is consistent with the hypothesis that quasars are not local objects. It is a more formidable task, however, to demonstrate that high-redshift quasars are at their extreme cosmological distances.

Until very recently the number of discovered lenses has been limited to $z_L < 1$. Strong lensing galaxies become increasingly rare at higher redshifts. The density of sources behind potential lenses is greatly reduced, and those with the highest lensing probability are usually red and faint in the optical surveys. Their confirmation requires more challenging near-infrared spectroscopy. Van der Wel *et al.* (2013) broke the $z_L = 1$ barrier using imaging from HST and spectroscopy from the Large Binocular Telescope.[18] Their remarkable discovery of the quadruple galaxy-galaxy lens J1000+0221 has been spectroscopically confirmed at redshifts of

$z_L = 1.53 \pm 0.09$ and $z_S = 3.417 \pm 0.001$. Figure 2.3 shows the almost continuous Einstein ring with a radius of ~0.35 arcsec. HST photometry of the source object shows a very blue continuum, indicative of a young, dust-free galaxy.* Soon after this famous discovery, Wong *et al.* (2014) set the new distance record with the discovery of a lensing galaxy at $z_L = 1.62$.[19] The search continues for a quasar source lensed by a galaxy at $z_L > 1$.

The universe does not lend a large percentage of its objects to gravitational lensing, and less than 1% of lensed objects are AGNs. The CASTLES Survey (**C**f**A**-Arizona **S**pace **T**elescope **LE**ns **S**urvey) website provides extensive reporting of all discovered lens systems.[20] Its compilation makes clear that very little data has been gathered for quasars lensed by galaxies at $z_L > 0.5$.

We are led to conclude:

1) Gravitational Lensing can disprove the local hypothesis (LH) of AGN redshift.
2) Gravitational Lensing cannot disprove the variable hypothesis (VH) of AGN redshift.†

Lyman-Alpha Absorption

Spectroscopy is an essential ingredient in astronomical exploration. By means of Doppler shift measurements, we can ascertain many galactic object properties: distance, temperature, density, mass, chemical composition, luminosity and relative motion. We assume the accuracy of the measurements based on cosmological redshift. Our understanding of deep space and cosmology is thus dependent upon the validity of CH. Neither 'near neighbor' association nor gravitational lensing validates it with certainty. It is therefore critical to understand how spectroscopy may do so.

Absorption lines in the spectra of galaxies and quasars are formed when electrons transition between energy levels. Neutral hydrogen atom electrons transitioning between ground state and higher energy levels produce the Lyman series of spectral lines. The electron transition between ground state and first excited state in the hydrogen atom is known as the

---

* There remains the possibility that two separate source galaxies are actually gravitationally lensed.

† Techniques of timing delay between the different lens images promise more robust measurements.

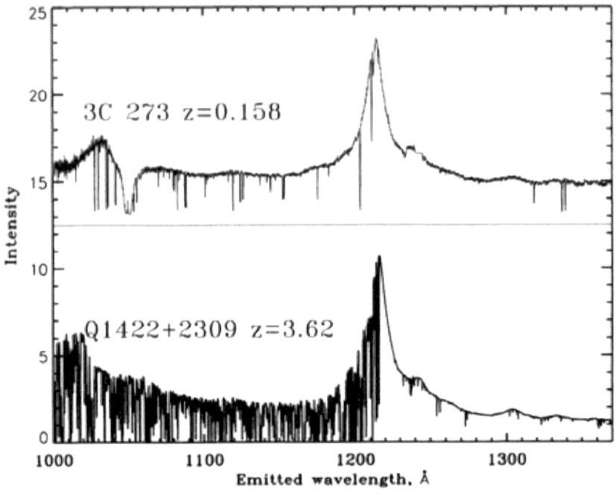

Figure 2.4 – Lyman-alpha absorption spectra of low and high redshift QSOs (shifted to common wavelength scale)

Lyman-alpha transition. The associated absorption line is known as the Lyman-alpha (Lyα) spectral line. Lyα lines have a rest wavelength of 1216 Å, which is in the ultraviolet portion of the electromagnetic spectrum. Neutral hydrogen (H_I) clouds in the intergalactic medium along the line of sight will produce Lyα absorption spectral lines observed throughout a range of wavelengths. The redshifts, z_{abs}, of these absorption lines will reflect the cosmological distances of the absorbing clouds. Since the light originates from the QSO at z_{em}, we see the QSO spectra with absorption lines characterized by $z_{abs} < z_{em}$ (see Figure 2.4). An observable thick series of Lyα lines is known as a Lyman-alpha forest (LAF). See Q1422+2309 in Figure 2.4. The first LAF was discovered in 1970 in an observation of the quasar 4C05.34 ($z = 2.877$). Observers differed in their explanation of the forest. Some argued that the absorbing gas is ejected from the QSO (the *intrinsic* hypothesis), while others argued the absorbing gas is cosmologically distributed (the *intervening* hypothesis). Within ten years evidence was offered to support the *intervening* IGM absorption theory. In turn, this provided partial evidence to support the measured redshift distance of the QSO and indirect support of the cosmological hypothesis. Verification was achieved by two different methods: statistical analysis and correlation to observable intergalactic galaxies and clouds.

The intervening hypothesis predicts that the number of Lyα absorption lines per unit redshift $N(z_{abs})$ must be identical for any QSO with $z_{abs} < z_{em}$. The large number of Lyα lines in an LAF provides opportunity for a statistical analysis of the hypothesis. Sargent, Young, Boksenberg, and Tytler (1980)[21] analyzed the Lyα absorption lines ($1.7 < z_{abs} < 3.3$) in high-resolution spectra of five QSOs: PKS 2126-158, Q0002-422, PHL 957, Q0453-423, and B2 1225+31.7. The observations were at ~0.8 Å resolution over a wavelength range of at least 1500 Å in the Earth's frame and with the Lyα emission line roughly centered within this range. They subjected the samples to the following statistical tests:

1. Comparison of the mean number density of Lyα lines among different QSOs
2. Comparison of the Lyα line density in different spectral regions in each QSO
3. Comparison of the mean rest equivalent width of the Lyα lines among different QSOs
4. Comparison of the mean rest equivalent width of Lyα lines in different spectral regions
5. Evaluation of the distribution in ejection velocity
6. Evaluation of the two-point correlation function in redshift among Lyα lines in each QSO

Each test was *entirely consistent* with the intervening hypothesis, and the Lyα lines are *very unlikely* due to material ejected from the QSOs themselves. Ejection was not completely ruled out because undiscovered factors *may possibly* make $N(z_{abs})$ the same for both the intervening and intrinsic hypotheses. They concluded that "the metal line systems probably arise in galaxy halos, and the Lyα clouds must be an intergalactic population which is not associated with galaxies." There are three noteworthy points to be made concerning the work of Sargent *et al.* (1980). First, although the sample of spectral lines is large, the number of QSO objects is limited to five,* all in the range $2.2 < z_{em} < 3.3$. Second, ejection theory is the opposing hypothesis that is under consideration (the theory of intrinsic gravitational redshift is not considered). Third, no basis is found for the claim by Bradley M. Peterson that the QSO with the larger redshift has the greater number of absorption lines. It is surprising in retrospect that Peterson's erroneous claim is never mentioned in later literature.

The most commonly detected spectral absorption lines are Lyαλ1216, the Mg_{II}λλ2795,2803 doublet, and the C_{IV}λλ1548,1550 doublet. It became quite natural for Sargent, Young and Boksenberg to go forward with research on the Mg_{II} and C_{IV} metal absorption lines. With 2.5 Å resolution observations from an additional 27 QSOs at redshifts $z_{em} \sim 2$, they found

* Statistical analysis based on a sampling of five objects would be deemed inadequate by today's standards.

evidence consistent with their previous work: (1) $C_{IV}\lambda\lambda1548,1550$ lines have a random redshift distribution consistent with absorption from intervening galaxies; (2) $Mg_{II}\lambda\lambda2795,2803$ absorption is not common, but its occurrence appears to be consistent with the predictions of randomly distributed intervening galaxies in a Friedman universe.[22]

The detection of K, Ca_{II} and 21-cm H_I in the spectrum of QSO 3C 232 arising from the halo of NGC 3067 offered hope for other absorption analyses. However, Weymann *et al.* (1981) documented many attempts that failed to find such intervening systems.[23] For example, close inspection of Markarian 205, which lies between a spiral arm and nucleus of NGC 4319, did not show any Ca_{II} absorption. Furthermore, none of the QSO-galaxy pairs examined by Haschick and Burke (1975)[24] showed any such absorption at a redshift that can be attributed to deflecting galaxies or clusters, and the research of Peterson (1978)[25] for 21-cm absorption in QSOs projected against large clusters yielded negative results. Subsequent statistical analyses did not provide any additional insight into these alternate QSO absorption lines.

A different statistical approach was taken by Richards *et al.* (1999)[26] in a work that compares intrinsic QSO properties against C_{IV} systems whose absorption redshifts are no greater than 71,400 km s^{-1} ($\beta = v/c = 0.238$).* The QSO intrinsic properties under comparison were absolute optical magnitude, redshift, radio spectral index, and radio morphology. The intervening hypothesis is deemed invalid if/when the C_{IV} absorption shows significant dependence on any property. The first property, absolute magnitude, is divided into 'Bright' and 'Faint' relative to the median absolute magnitude of the sample data, $M_V = -27.0$. Figure 2.5 shows a notable excess of absorbers, N(β), in bright QSOs over faint QSOs from 15,000 to 65,000 km s^{-1} ($0.05 < \beta < 0.2167$), with a peak near $\beta = 0.1$c. This observed dependence of C_{IV} absorption on absolute magnitude is inconsistent with the intervening hypothesis.

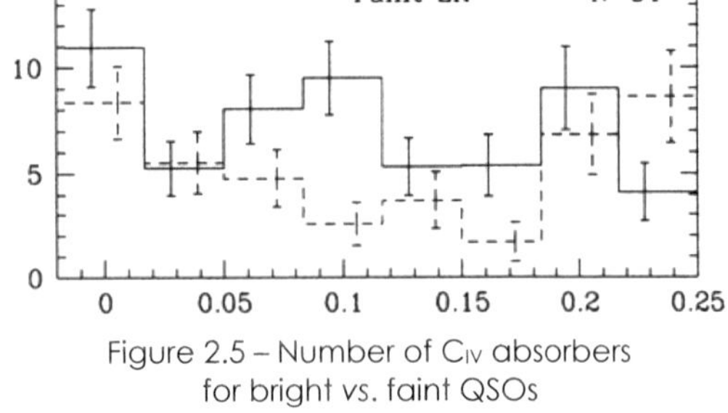

Figure 2.5 – Number of C_{IV} absorbers for bright vs. faint QSOs

* Systems with larger β become blended with the Lyα forest.

Other graphs in their paper show similar dependencies of the absorber count on the selected intrinsic property. This suggests that *some* of the absorption is intrinsic to the QSO. They concluded that "studies of absorption-line systems will have to be reconsidered."

If Weymann *et al.* (1981) and Richards *et al.* (1999) cast any doubt on the landmark findings of Sargent *et al.* (1980), the alternate statistical analysis of Libby *et al.* (1984)[27] may undermine it completely. Libby's analysis uses 13 QSOs (previously measured by Roberts *et al.* (1978), Young *et al.* (1979), Sargent *et al.* (1979), Sargent *et al.* (1982), and Young *et al.* (1982)[28][29][30][31][22]) with an average of 73 absorption lines each. The wavelengths are said to 'concur' from QSO to QSO if they lie within a wavelength interval of 1.0 Å. By such concurrence 55 of the absorption lines are found in at least 5 of the 13 QSOs. This compares with only 13 ± 2 quintuple concurrences that are found by random coincidence. This difference in number is statistically significant. The strong implication is that *the absorption lines are not principally caused by absorption in clouds randomly distributed between each QSO and the Earth.* Exception to this analysis is taken in another paper (Varshni & Singh, 1985) that shows a less significant random coincidence value of 38.5 ± 3.8.[32] An error in the computer random number generator is the suggested explanation for Libby's smaller value. The strength of any numerical analysis must remain in some doubt if underlying computer algorithms are called into question.

Stronger evidence will come from actual observations of any intervening formations. The more foreground objects found whose redshifts match the QSO absorption lines, the stronger is the argument for the intervening hypothesis. Bergeron and Boisse (1991) documented the first sample of galaxies that gives rise to Mg_{II} absorption line systems in quasar spectra. The intervening absorbing galaxies are at $z < 0.5$ and typically lie at an angular separation in the range 5 to 12 arcsec, as shown in Table 2.2.[33]

In their survey LeBrun *et al.* (1993) conclude that "for 13 among the 14 confirmed Mg_{II} systems there is a bright absorber candidate at a linear separation less than $4R_H^*$ in the absolute magnitude range $-23.0 < M_r(z_a) < -19.5$. The probability that this occurs by chance for all the systems is very low, $P = 0.13\%$."[34] Together with the statistical arguments of Sargent *et al.*

* R_H is the (Holmberg) radius at which the surface brightness is 26.5 magnitudes per square arc second in blue light. ($R_H = 22\ h_{50}^{-1}$ kpc).

Galaxy	z_{QSO}	$z_{absorber}$	z_{galaxy}	Angular separation in arcsec	Apparent magnitude (r)	Absolute magnitude M(r)
0109+200	0.746	0.5346	0.534	7.1	21.7	−21.3
0151+045	0.404	0.1602	0.160	6.4	19.1	−21.0
			0.160	10.9	20.2	−19.9
0229+131	2.065	0.4176	0.417	6.8	20.05	−21.9
0235+164	0.940	0.524	0.524	1.9	21.2	−21.8
0453–423	2.661	0.7261	0.726	30.3	22.3	−21.6
1038+064	1.270	0.442	0.441	9.4	21.2	−21.4
1101–264	2.148	0.3591	0.359	12.2	20.4	−21.6
1127–145	1.187	0.3130	0.313	9.6	19.5	−22.2
1209+107	2.191	0.3930	0.392	7.1	21.9	−20.3
1332+552	1.24	0.374	0.373	5.0	20.7	−21.3
1511+103	1.546	0.4369	0.437	6.9	21.6	−20.9
2128–123	0.501	0.4299	0.430	8.6	21.0	−21.5
2145+067	0.990	0.7908	0.790	5.5	22.5	−21.6

Table 2.2 – Mg_{II} absorbing galaxies

(1980), this led to the general view that the absorbers are individual clouds lying at smaller distances than the QSO.

Although they caused a shift in opinion toward the intervening hypothesis for QSO spectral line absorption, these studies did not fully substantiate the cosmological hypothesis. Mg_{II} absorption at $z_{abs} < 0.5$ is only one type of spectral line in a small fraction of the universe. Evidence must at least explain the density of Ly*a* absorption lines at high redshift (see Figure 2.4).

If intervening clouds indeed cause the LAF, they must be shown to exist. Such evidence is difficult to obtain. The typically weak QSO absorption lines require high spectral resolution available only on the world's largest telescopes. Any line at high redshift requires observation from the rest wavelength λ_0 to the longest wavelength, $\lambda_0(1 + z_{em})$. This necessitates further observations from ultraviolet satellites. Since the launch of HST, we know that Lyα lines persist to zero redshift.[35] A study of the Lyα absorbing systems at smaller redshifts is the best way to approach the more difficult observations. Lanzetta *et al.* (1995) considered the origin of Lyα absorption at redshifts spanning $z = 0.0700\text{-}0.5526$ and reached the following conclusions:[36]

1) At $z < 1$ most luminous galaxies are surrounded by extended gaseous envelopes of radius ~160 h^{-1} kpc and of roughly unit cover factor.
2) At $z < 1$ the fraction of Lyα absorption systems, including Lyα-forest absorption systems, that arises in luminous galaxies is at least 0.32 ± 0.10 and may be as high as 0.60 ± 0.19.

These are not the desired conclusions that bolster CH. First, it is difficult to conceive of the galactic haloes extending to such a distance. It requires an *ad hoc* explanation (more dark matter?). Second, the high percentage of luminous galaxy absorbers runs contrary to Sargent *et al.* (1980), who stated that Lyα forest absorption systems arise instead in intergalactic clouds. Nevertheless, the intervening hypothesis is strengthened for low redshift absorption.

Number	Quasar	m_V (mag)	m_H (mag)	z_{em}	z_{abs}
1	[HB89] 0000–263	18.0	15.8	4.10	3.40
2	LBQS 0010–0012	18.0	17.4	2.14	2.03
3	LBQS 0013–0029	17.0	16.0	2.08	1.97
4	LBQS 0102–2014	18.0	16.9	1.98	1.74
5	[HB89] 0249–222	18.4	16.2	3.20	2.83
6	[HB89] 0454+039	16.5	15.4	1.35	0.860
7	[HB89] 0809+483	17.8	16.3	0.871	0.437
8	[HB89] 0836+113	18.8	17.8	2.70	2.47
9	[HB89] 1151+068	18.8	16.6	2.76	1.77
10	[HB89] 1157+014	17.7	15.7	1.99	1.94
11	[HB89] 1209+093	18.5	17.4	3.30	2.58
12	[HB89] 1215+333	17.5	17.1	2.61	1.99
13	BRI 1346–0322	18.8	17.0	3.99	3.73
14	BRI 1500+0824	18.8	17.6	3.94	2.80
15	Q1610+2806	19.2	18.2	3.51	2.59
16	JVAS 1757+7539	16.1	15.2	3.05	2.63
17	LBQS 2206–1958	17.3	15.7	2.56	1.92
18	LBQS 2206–1958	17.3	15.7	2.56	2.08
19	[HB89] 2348–011	18.0	17.0	3.01	2.43
20	[HB89] 2348–011	18.0	17.0	3.01	2.61
21	LBQS 2359–0216	18.0	17.3	2.82	2.10
22	LBQS 2359–0216	18.0	17.3	2.82	2.16

Table 2.3 – Quasars with DLA system

Colbert and Malkan (2002) considered the more problematic high-z Lyα absorption.[37] They imaged 19 quasars with 22 damped Lyα systems (DLAs) using the F160W filter and the Near-Infrared Camera and Multi-Object Spectrograph on board the HST. Their sample contains a median redshift of $z = 2.10$ (see Table 2.3). *No galaxies were seen near quasar lines of sight that contain the DLA absorption* despite the ability to look much closer, 0.5 arcsec or better, to the quasar than other deep imaging surveys. What possible (*ad hoc*) explanation is there for Lyα absorptions by galaxies at $z < 1$ when there are none found at high redshift? They were left to conclude: "From the evidence gathered so far, it appears that the distribution of DLAs is inconsistent with the evolution of galaxies."

We are led to conclude:

1) Lyα Absorption can disprove the local hypothesis (LH) of AGN redshift.
2) Lyα Absorption cannot disprove the variable hypothesis (VH) of AGN redshift.

Summary

The three pillars that support the cosmological redshift hypothesis (CH) are the *near neighbor* relationship, gravitational lensing, and LAF spectra. The *near neighbor* argument is weakened by discoveries of galaxies near higher redshifted quasars. These may be overlooked as statistical flukes. The gravitational lensing argument is weakened by the paucity of observations and the current lens limitation of $z_{lens} < 0.5$. This also may be overlooked given how few opportunities exist for proper lens alignment. The spectroscopy argument is weakened by the fact that Lyman-alpha forests are not always found for some quasars. Such observations are not always possible, and it is reasonable to assume the LAFs will be found with future technology. As a whole, the evidence does make an overpowering case that quasars are not local objects, as once believed by some astronomers.

The technical limitation of high-z observations makes it impossible to confirm the universality of CH. Is it then reasonable to assume that high-z quasars behave in similar fashion to those at closer distances? The answer would probably be 'yes' if observations were limited to the three pillars in support of CH. But there is more to the story. The next chapter reveals numerous other observations that suggest something more complex is happening in the universe.

Chapter 3

Difficulties with the Cosmological Hypothesis

Absent any conclusive evidence of the cosmological hypothesis (CH), some researchers continue to sprinkle the journals with articles that call its tenets into question. To appreciate the persistent doubts requires an understanding of the implausible intrinsic magnitude that CH implies for quasars. The maximum luminosity of an object powered by accretion is reached when the force of radiation acting outward becomes equal to the gravitational force acting inward. This is known as the Eddington limit, or Eddington luminosity. Furthermore, the radiation must be produced in a region constrained by $R < c\Delta t$, where Δt is the time that light takes to cross the emitting region. It can be shown that the luminosity of a maximally efficient accreting source is $L_{max} \sim \tau 10^{41}$ ergs/sec, where τ is the variability time in seconds.* The problems for CH start with the observation that quasars with luminosities from 10^{46} to 10^{47} ergs/sec have been measured with a variability of a few hours. Their luminosity is at least two orders of magnitude larger than allowed. Further difficulties come from measurements of quasar age, variability, metallicity, dust, Faraday rotation, and more.

It is quite common to tweak a scientific theory when new observations make it necessary to do so. In the study of quasars surprising new observations come frequently, many of them difficult for CH to explain. Some theoreticians believe that some 'new physics' is necessary. After all, anything is possible when dealing with the laws of the unknown universe. The overarching issue, though, is not how many difficulties arise from the premise of the cosmological hypothesis. It is whether a different theory explains things better or whether CH stands in conflict with anything else. This is not easy to evaluate. It requires bringing all the difficulties with CH to the fore.

---

* See the section, **Variability**, in this chapter for further details.

Evolution

The data acquired from ($z < 3$) quasar surveys by end of the twentieth century indicated that quasar density increases with higher redshift for any given luminosity M_B. This implies that there were more quasars or brighter quasars in earlier cosmic history or, more likely, some combination of both. It is also interesting that no significant change is found in their radii, morphology distribution, or spectra.* A tool often used to depict quasar evolution is the Quasar Luminosity Function. The QLF is depicted in Figure 3.1. The space density, $\phi(M_B..z)$, is the number of quasars per unit comoving volume of the universe.[38]

Comoving density is used instead of proper density in order to counterbalance the changes that arise from the expansion of the universe.† It is an outstanding problem to formulate an astrophysical theory that predicts both the origin and QLF evolution of quasars. The stream of data coming from subsequent space surveys makes the problem evermore difficult to resolve.

The most popular cosmology today is ΛCDM (Lamda Cold Dark Matter), a derivative of the Big Bang theory. In this cosmology the universe is 13.7 billion years old and began its period of large-scale structure formation about 150 million years after its origin. This implies a finite amount of time for any structure to evolve. Consequently it is difficult for Big Bang models to explain the large structures observed in the very early periods of the universe. Most difficult are early quasars with a mass of 10^9 M_{Sun} that took less than 600 million years to evolve. Consider a starting mass of $M_{Seed} = 10$ M_{Sun}, a typical radiation efficiency of 10%, and accretion at the Eddington limit. After 600 million years, the mass of the black hole has

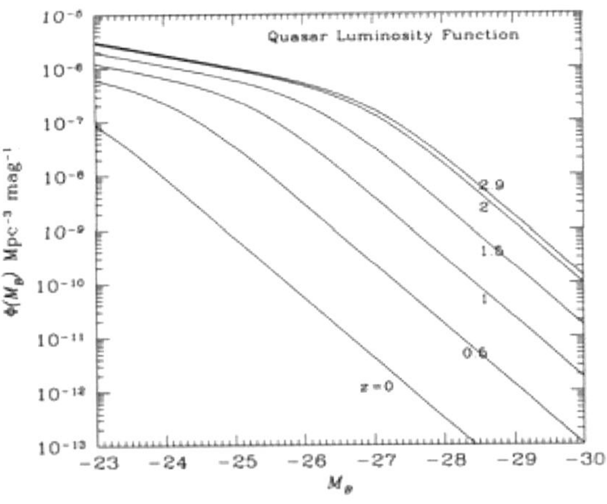

Figure 3.1

* However, see Chapter 6 for discussion of increased spectral blueshifts as a function of z.

† Proper_Volume is proportional to (Comoving_Volume)·$(1 + z)^{-3}$

a size of $M_{BH} = 8.5 \times 10^6\ M_{Sun}$. This is short of the required ($10^9\ M_{Sun}$) mass by *two orders of magnitude.*

The turn of the century saw growing concern about 'over-massive' quasars. First came the discovery of the $z = 6.30$ QSO SDSS J1030+0524, whose appearance was indistinguishable in any way from lower redshift QSOs. This quasar would need a starting mass of $M_{Seed} = 100\ M_{Sun}$ and grow at the Eddington-limited rate for ~725 Myr. Farrah *et al.* (2004) found little likelihood for such a scenario and declared, "accretion onto an initially stellar mass black hole is unlikely to produce the central black hole in SDSS J1030+0524 within the required timescale, and a significant fraction of the black hole mass must be built *via* some other mechanism." [39] Within a couple of years nine additional quasars with apparent magnitude of ~19 were discovered at $z > 6$. By then it became certain that the strength of QSO emission lines remains roughly constant throughout history. It also became certain that well-evolved black holes exist that had insufficient time to grow even given the unlikely scenario of continuous Eddington accretion rates. [40] In addition to their large mass, these $z > 6$ QSOs have strong and rapid chemical enrichment, placing further tight constraints on theoretical models. Their progenitors must be lower mass black hole seeds with some unknown capability for rapid evolution not seen in local galaxies. Djorgovski *et al.* (2008) studied possible formation mechanisms for these objects. They suggest that the best possibility is stellar mass black holes produced as endpoints of massive Population III star evolution, but "it is not obvious how such vigorous accretion could have taken place." [41] The problem becomes more difficult if these are *spinning* black holes. Radiative efficiency for spinning BHs would make the necessary evolution period approach 2 Gyr. Adding further mystery was the discovery in 2015 of J0100+2802 at $z = 6.30$ with a $1.2 \times 10^{10}\ M_{Sun}$ black hole. [42] The discovery of the Ultraluminous Quasar further spoiled the efforts to explain QSO evolution. The $M_{Seed} = 10\ M_{Sun}$ quasar now falls short of the required ($10^{10}\ M_{Sun}$) mass by *three orders of magnitude.*

If quasars are too large in the early universe, then they are also too small in the current epoch. The statistics of the Sloan Digital Sky Survey (SDSS) survey show that the masses of the observed central BHs have decreased from $< 10^{9.5}\ M_{Sun}$ to the current $< 10^7\ M_{Sun}$. [43] If super-massive black holes existed ten billion years ago, the same or larger holes should exist today. Hawking radiation is the only known mechanism for black holes to lose their mass.[44] However, the rate of evaporation is too small. It would take ~10^{20} years before black holes begin evaporating faster than the rate of growth. That leaves a picture of a universe that created objects that are too large and later disappear without any physical explanation. Is it possible that this 'counter-evolution' is a mirage caused by faulty distance calculations?

Iron Abundance

The remarkable evolution of AGN luminosity from early epochs to the present would suggest features that should change through the eons. An obvious candidate for change is the metallicity. We know that young stars are mostly devoid of heavier elements, and we may expect a similar dearth in the young AGNs. Research has demonstrated this is not the case. The youngest AGNs show element counts at least as high as those of the modern epoch. Dietrich *et al.* (2003) found a trend of increasing metallicity with luminosity and several times solar metallicity in quasars at $z > 3.5$,[45] and Fan *et al.* (2001) detected line ratios $N_V\lambda1240/C_{IV}\lambda1549$ and

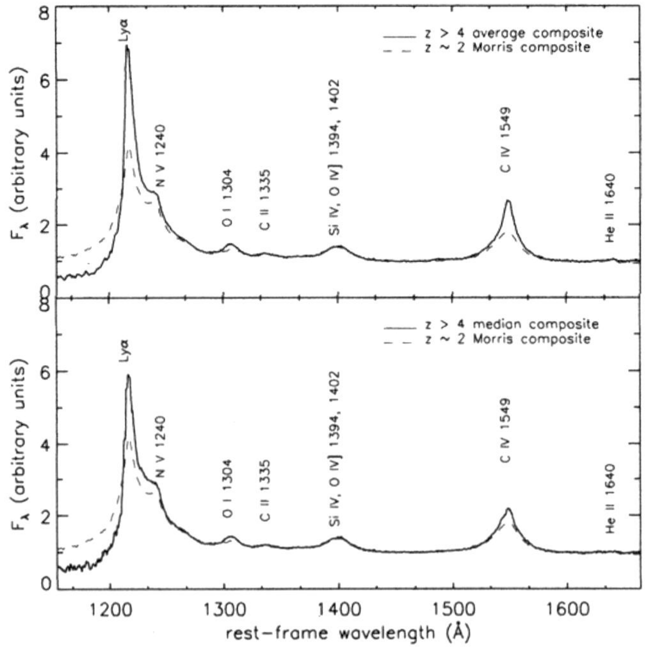

Figure 3.2 – Quasar emission lines

$N_V\lambda1240/He_{II}\lambda1640$ that demonstrate supersolar metallicity in the BELR region of QSOs at $z \sim 6$.[46]

Constantine *et al.* (2002) studied the emission line properties of quasars at $z > 4$ *vs.* $z \sim 2$. C_{II} and Si_{IV} are basically unchanged, while N_V, C_{IV} and O_I have a higher incidence in the $z > 4$ QSOs.* The results are shown in Figure 3.2.[47] The strength of N_V emission implies characteristically higher metallicities in the $z > 4$ objects. Given the large amount of time necessary for evolution of the QSO and underlying host galaxy, we now must assume substantial star formation preceding QSO existence. Thompson *et al.* (1999) showed a similar increase in the iron abundance in QSOs out to $z = 4$.[48] The abundance of Fe_{II} in high-z QSOs (exceeding levels found in QSOs at lower redshift) puts unique constraints on luminous QSO broad-line physics and nucleosynthesis in the QSO host galaxies.

Most iron is created in Type Ia supernovae from progenitors with long evolutionary timescales, whereas magnesium is produced on rapid timescales by SN Types II, Ib, and Ic. We thus expect to see dramatic evolution of the Fe_{II}/Mg_{II} line ratio after the first SN Ia progenitor population explodes and enriches the interstellar medium. Maiolino *et al.* (2003) investigated this ratio in the near-IR spectrum of quasar SDSS J114816.64+525150.3 ($z = 6.4$) and found it larger than all but two in their sample of 22 quasars ($3 < z < 6.4$).[49] This implies nearly constant or even slightly increasing iron abundance from $z = 0$ to $z = 6.4$ (800 Myr after the Big Bang). The finding is difficult to reconcile with scenarios expecting a delay of at least 1 Gyr for iron enrichment. Barth *et al.* (2003) further estimated that the SN Ia progenitor stars are constrained to form at $z > 10$.[50] Iwamuro *et al.* (2002) studied the

Median Values of Fe_{II}/Mg_{II}

Redshift	Fe_{II}/Mg_{II}	Error	Number
0101 ± 0068	$4.48^{+4.94}_{-1.69}$	0.84	30
0836 ± 0086	$3.34^{+1.95}_{-0.87}$	0.80	225
1010 ± 0088	$3.62^{+2.02}_{-1.04}$	0.75	225
1169 ± 0070	$3.49^{+2.16}_{-0.93}$	0.77	225
1303 ± 0065	$3.02^{+2.16}_{-0.88}$	0.78	225
1441 ± 0072	$2.68^{+1.80}_{-0.74}$	0.80	225
1561 ± 0048	$2.91^{+1.78}_{-0.86}$	0.79	225
1664 ± 0054	$2.80^{+1.80}_{-0.74}$	0.92	225
1774 ± 0056	$2.99^{+2.23}_{-0.93}$	0.93	225
1917 ± 0085	$3.92^{+4.15}_{-1.30}$	1.37	225
2146 ± 0143	$5.14^{+6.96}_{-1.94}$	1.98	218
3411 ± 0220	$5.50^{+6.77}_{-2.99}$	1.46	6
4475 ± 0215	$4.40^{+7.22}_{-0.61}$	1.74	9
4997 ± 0291	$5.69^{+4.25}_{-3.71}$	2.70	8

Table 3.1 – Fe_{II}/Mg_{II} line ratios

* At $z > 4$ the Lyα emission is considerably affected by the forest absorption.

Fe_{II}/Mg_{II} ratios of several thousand quasars. Table 3.1 shows an *increasing* trend in the Fe_{II}/Mg_{II} ratio from $z = 1.5$ to $z \sim 5$, independent of luminosities. [51]

Freudling *et al.* (2003) further explored QSOs in a combined sample of comparable luminosities. They concluded that, "There is no compelling evidence of evolution in QSO FeII/MgII abundance from $z \approx 0.1$ to $z \approx 6$." [52] Their findings severely constrain current cosmological models with more *ad hoc* assumptions. The constraint would diminish proportionately if some intrinsic redshift were ascribable to the QSOs.

X-Ray Spectrum

If no evolution exists in the metal content of QSOs, then perhaps some subtle change occurs in the propagation of the electromagnetic waves. Spectroscopy is the ideal means to find such changes. Spectra in every wavelength now exist for thousands of objects at every cosmological distance. It has been determined that the X-ray band is best for the study of AGNs. X-ray emissions are a universal characteristic of all AGNs, and X-ray luminosity appears to fade less than optical. Some uncertain characteristic process of evolution seems to make X-ray fluxes stronger than optical fluxes as quasars become less luminous. The X-ray continuum in the 2-10 keV energy range (soft X-rays) is best for analysis and can usually be approximated by the statistical relation $L_{\text{optical}}^{0.7} \sim L_{\text{X-ray}}$. For energies less than 1 keV, lower luminosity sources can be affected by absorption along the line of sight; for energies greater than 10 keV, there can exist a flattening of the spectrum. X-rays make a good tracer of AGN activity. The samples are complete, uniform and less affected by obscuration at high energies. Once an object is located in the X-ray spectrum, the redshift then is determined by optical and infrared analysis.

Plotting photon flux versus wavelength is a common method to study a limited range of the electromagnetic spectrum. However, spectra are often of power-law form and require the construction of the Spectral Energy Distribution (SED) for more detailed analysis. The number of photons in an energy band per unit time and unit area then follows the law $n(E) = n_0 E^{\tau}$, where n_0 is the normalization of the power law and τ is the *photon index.** It is this photon index τ that is often depicted in graphs that

* The opposite sign convention is sometimes employed such that $n(E) = n_0 E^{-\tau}$.

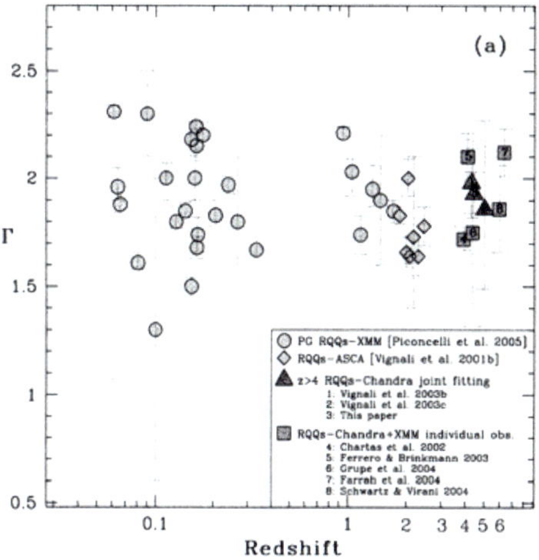

Figure 3.3 – X-ray photon index vs. redshift

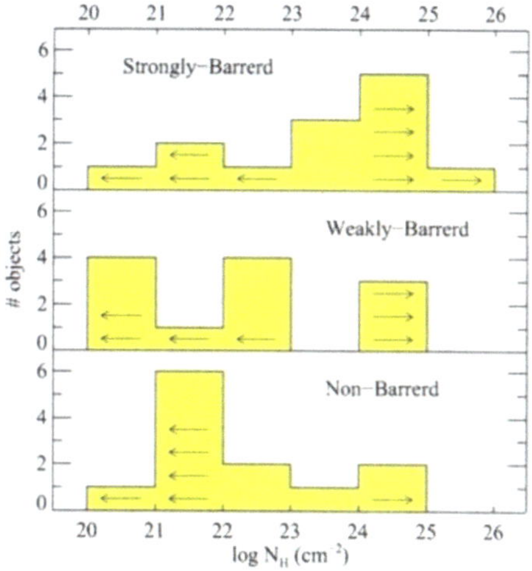

Figure 3.4 – Column density distribution (three strengths of host galaxy stellar bar)

provide spectral features. The energy flux is then the energy of the photons multiplied by the photon flux, $\phi(E) = E^*n(E) = n_0E^{\tau+1}$.

Reeves and Turner (2000) drew on the database of all available ASCA* observations through January, 1998, in order to correlate the X-ray photon index τ against radio-loudness R_L, X-ray luminosity L_x, and quasar redshift z.[53] They limited their analysis to the 68 objects with $z > 0.05$ and with $M_V < -23.0$. A strong correlation was shown between τ and radio-loudness R_L. No significant correlations were found between τ and either L_x or z for the radio-quiet quasars in the sample. Radio-loud quasars were discounted in the latter analysis due to the complication of powerful relativistic radio-jets that contaminate the data. The only possible evolution in the quasars was in the amount of neutral X-ray absorption. It was left open whether this absorption is intrinsic to the quasars or originates from line-of-sight matter.

Vignali *et al.* (2005) published similar results for quasars out to $z = 6$.[54] Figure 3.3 shows there exists a scatter of $1.5 < \tau < 2.5$ for the quasars taken from all surveys, but no detectable change in the photon index occurs as a

* Advanced Satellite for Cosmology and Astrophysics – the fourth X-ray astronomy mission by Japan and the first to use CCDs for X-ray imaging with a broad bandpass, good spectral resolution, and large effective area.

function of redshift. This suggests that the physical process of accretion onto black holes has not changed over cosmic time. This is quite unexpected given the change in environment and number density that has occurred. They infer that "X-ray emission regions seem to evolve relatively little from $z = 0$ to $z \sim 6$."

A further approach to the study of radiation absorption concerns the phenomenon of Compton scattering. When the column density distribution is $10^{24} < N_H < 10^{25}$ cm^{-2}, the gas is thick to Compton scattering ('Compton thick'), rendering energies below 10 keV invisible. When $N_H > 10^{25}$ cm^{-2}, X-ray radiation in the range 10-100 keV is also lost. Maiolino and Risoliti (2007) showed that the Compton thick AGNs were at the same relative fraction locally as those at high-z.[55] Although they found no evidence of evolution in X-ray absorption, they did find an unexpected correlation between absorption and the bar strength of the host galaxy. Figure 3.4 shows that most Compton thick nuclei are hosted in barred galaxies. The galaxy bar exists on the kpc scale. It is difficult to explain how the bar affects the Compton thick region which must be located within the central ~10 pc. If X-ray absorption evolves with galaxy morphology, it makes it more difficult to explain the lack of evolution over cosmic time.

Dust

An important component of the interstellar medium is the solid grains of material known as interstellar dust. These dust grains consist of compounds with some (element) subset of {C, O, Mg, Si, Fe} with radial sizes of 0.001-0.1 μm. Dust is produced by supernovae and in the envelopes of evolved (> 1 Gyr) stars and is expelled by supernova explosions and stellar winds. Its presence is detected by background light attenuation known as interstellar extinction. Short-wavelength emission is attenuated more than long-wavelength emission, thereby causing a reddening of the background source. It is possible to compare the observed color/brightness *vs.* intrinsic color/brightness to deduce the amount of dust along the line of sight. Submillimeter observations are ideal for such observations.

Archibald *et al.* (2001) presented results of the first major systematic submillimeter survey of radio galaxies spanning the redshift range $1 < z < 5$.[56] They found, "clear evidence that the typical submillimeter luminosity (and hence dust mass) of a powerful radio galaxy is a strongly increasing function of redshift, rising at the rate $(1 + z)^3$ out to $z \sim 4$." While this result does not violate CH, it does require any AGN evolution model to predict decreased activity. The real problem came from a report (Bertoldi *et al.*, 2003) which studied dust emission from quasars at $z > 6$.[57] It reveals dust masses of $4 \times 10^8\ M_{\text{Sun}}$ for J1048+4637 ($z = 6.23$) and $7 \times 10^8\ M_{\text{Sun}}$ for J1148+5251 ($z = 6.42$). At constant formation this implies an extremely rapid dust production rate of $\sim 1\ M_{\text{Sun}}\ \text{yr}^{-1}$. It also implies a high abundance of heavy elements less than 1 Gyr after the Big Bang.* Dunne *et al.* (2003) considered supernovae as an explanation of the high interstellar dust.[58] They estimated that the current Galactic dust production rate from Type II supernovae is $7\text{-}18 \times 10^{-3}\ M_{\text{Sun}}\ \text{yr}^{-1}$. When this is added to stellar wind sources of $5 \times 10^{-3}\ M_{\text{Sun}}\ \text{yr}^{-1}$, a twofold order of magnitude shortfall of the observed dust production remains in high-redshift QSOs. While adding dust from supernovae can account for another order of magnitude increase in total dust mass, supernovae shocks have the potential to evacuate the dust and further exacerbate the dust budget.

Rowlands *et al.* (2014) investigated the 'dust-budget-crisis' with best-fit analysis of chemical evolution.[59] The only model that fits observations requires raising dust emissivity† by a seven-fold factor and assuming dust is shielded in cold, dense regions of the ISM that are unaffected by supernovae shock waves. There currently is no observational evidence either for the large variations in dust emissivity or for any area shielded from shock waves.

Faraday Rotation

The Faraday effect is an interaction between light and a magnetic field that causes a rotation of the plane of polarization. The amount of rotation (β) in the polarization angle depends on the magnetic field strength ($B_{||}$), the electron density (n_e) along the line of sight, and the frequency of the electromagnetic wave. Interstellar Faraday rotation exhibits a simple dependence on the wavelength of light (λ) and rotation measure (RM) given by:

---

* See previous section, **Iron Abundance**, for other studies on high metallicity in high-redshift AGNs.

† Emissivity, κ, is the ratio of surface thermal radiation to the radiation from an ideal black surface at the same temperature as given by the Stefan-Boltzman law. At room temperature the light is infrared, invisible to the eye.

$$\beta = RM \cdot \lambda^2$$

$$RM = \frac{e^3}{2\pi \cdot m^2 c^4} \int_0^d n_e(s) \beta_{||}(s) ds.$$

Correlating redshift to the Faraday effect is not an easy task. The light from extragalactic sources must pass through an intergalactic medium in which the magnetic field will vary in strength and orientation. It is possible in some cases to have a net zero RM from highly redshifted objects due to magnetic field reversals. Nevertheless, arguments have been presented to support non-cosmological distances for QSOs based upon Faraday rotation.

Until the work of G.L. Welter, J.J. Perry, and P.P Kronberg (WPK), there was no way to know how redshift might correlate with Faraday rotation. Their 1984 publication [60] stated, "the *variance* of the RM distribution of QSOs at redshift z_e increases systematically with z_e (emission redshift), increasing by a factor of ~4 between $z_e = 0.4$ and $z_e = 2$." Figure 3.5 shows the tendency for RM values to be more tightly clustered around zero for small z_e. This is a most interesting claim. Redshift does not noticeably correlate with the absolute value of RM, but it does correlate with the *variance* of RM.

Their sample of QSOs also showed a strong correlation between the occurrence of 'excess' extragalactic Faraday rotation (more than ~30 rad m^{-2}) and strong absorption lines. They concluded: "The RM arises in a population of discrete intervening clouds although statistics are inadequate to rule out substantial intrinsic QSO contribution."

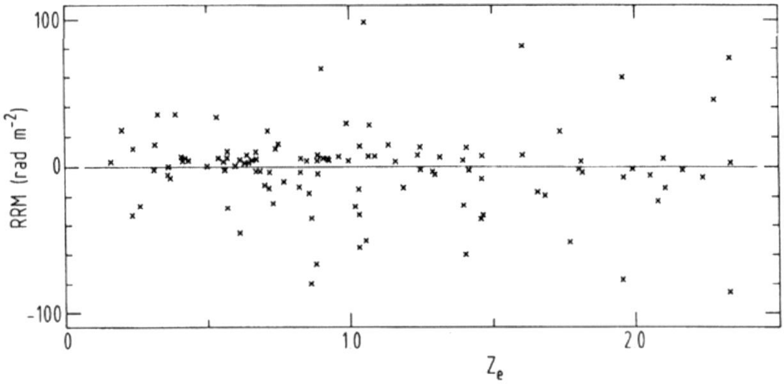

Figure 3.5 – RM vs. z_e for 112 QSOs

Coordinates (1)	Name (2)	RRM rad m^{-2} (3)	± ε rad m^{-2} (4)	z_e (5)	N_a (6)	$<z_e>$ (7)
0017+15	3C 9	–8	*	2.012	3	1.70
0119–04		–7	9	1.948	7	1.97
0122–00		9	7	1.080		
0232–04	OD 055	–2	3	1.434		
0237–23		–8	4	2.240	9	1.77
0350–07	3C 94	8	9	0.962		
0458–02	4C –02.19	100	*	2.28	1	2.28
0710+11	3C 175	–51	*	0.768		
0736–06	OI –161	–64	*	1.900	2	1.921
0802+10	3C 191	60	8	1.956	2	1.95
0805+04	4C 05.34	–35	*	2.877	7	2.50
0835+58	3C 205	–5	*	1.534	2	1.54
0836+19	4C 19.31	–20	5	1.691	1	1.42
0932+02	4C 02.27	–25	13	0.659		
0952+17	OK 106	9	9	1.472		
1104+16		8	11	0.634		
1137+66		3	5	0.652		
1148–00	OM –080	–1	2	1.982		
1218+33	3C 270.1	13	*	1.519		
1229–02	ON –049	–56	5	1.038	1	0.395
1317–00	4C 00.50	9	2	0.890		
1328+307	3C 286	3	1	0.849	1	0.692
1331+170		–23	6	2.081	5	1.776
1354+195	OP 191	2	*	0.720		
1421+12	OQ 135	9	6	1.604		
1442+101	OQ 172	21	9	3.53	4	2.78
1458+71	3C 309.1	66	8	0.904		
1510–089		–7	5	0.361		
1633+38	4C 38.41	–3	8	1.814		
1641+399	3C 345	0	4	0.595		
1828+48	3C380	–11	10	0.692		
2120+16	3C 432	–70	*	1.805		
2128–123		0	2	0.501		
2146–13	OX –178	–1	1	1.800	1	1.78
2223–05	3C 446	–27	5	1.403	1	0.847
2251+24	4C 24.61	–420	*	2.328	4	2.363
2251+158	3C 454.3	–19	17	0.859		
2344+09		10	3	0.677		
0957+561	QSO A	–63	10	1.405	1	1.39
0957+561	QSO B	–163	10	1.405	1	1.39

Table 3.2 - RRM of 39 QSOs

WPK compiled their data from many sources. A sample of 39 QSOs is shown in Table 3.2. Columns (3), (4), (5), (6) and (7) give, respectively, the residual rotation measure (RRM), its estimated error, the QSO emission redshift, the number of strong absorption systems detected, and the average absorption-line redshift for those objects with strong detected absorption. An asterisk indicates that the value should be used with caution. Noteworthy is the fact that those entries with a high number of strong absorption systems show an average emission redshift very close to the redshift of the QSO object.

Five years later Halton Arp explored these data in greater detail.[61] He found it strange that the RM for quasars of $z \sim 2$ was no greater, and probably even less, than for quasars of $z \sim 1$. He noted that the largest scatter of Faraday rotation occurs from broad absorption line quasars (BALs). ("A broad absorption line represents material close to the quasar that has been ejected from the quasar toward the observer.") BALs represent approximately 9% of the quasar population and are found most often at $z > 1.5$. This suggests that the scatter of Faraday rotation is an effect of morphology rather than distance, and that presents the question of why there would be no further increase in RRM beyond the redshift $z = 2$.

A further detail appeared when the data were divided between objects in the direction of the Local Group (Dec > –20°; 21^h < R.A. < 5^h) and objects in the direction of the Local Supercluster (Dec > –20°; 9^h < R.A. < 17^h). There are large numbers of low luminosity quasars in the direction of the Local Group at redshift $z \sim 0.5$ and $z \sim 2.0$. In the opposite direction, the Local Supercluster shows the predominance of high luminosity quasars at $z \sim 1$ whose RM values are essentially all negative. Since this negative bias does not occur in other directions, it suggests that the quasars are associated with the magnetized plasma of the Local Supercluster and not all at cosmological distances. The combined analysis of Arp and WPK adds one more reason to consider the possibility that not all AGN are at cosmological distances.

Variability

Another property that distinguishes AGNs from other galactic objects is the variation in their flux. AGNs are variable in every waveband in which they have been studied, both in the continuum as well as their broad emission lines. The especially strong X-ray variations make their energies ideal for quasar density surveys. Variability studies are important in identifying the size of the region in which the radiation is produced. If variability is observed on a time scale of Δt_{var}, then the radiation must be produced in a region with size constraint of $R < c\Delta t_{var}$. The X-ray variability is highly correlated to optical and UV continua whose spectra then are used for object identification. Luminosity variations of three magnitudes twice a day have been observed.

The stretching of time by a factor of $(1 + z)$ is a fundamental property of an expanding universe. Time dilation can thus be used to test the distance between a redshifted object and observer. Goldhaber *et al.* (2001)[62] studied the light curve widths of supernovae over a wide range of redshifts and found convincing evidence for cosmological time dilation.* M.R.S. Hawkins decided to extend this idea to the study of quasars. In his 2010 paper[63], "On Time Dilation In Quasar Light Curves," Hawkins compared time-scales of variation in samples of quasars at different redshifts. Quantification of time scales is provided well by the Fourier power spectrum,

$$P(s_i) = \frac{\tau}{N}\left[\sum_{j=1,N} m\left(t_j\right)\cos\frac{2\pi \cdot ij}{N}\right]^2,$$

where i runs over the N equally spaced epochs of observation separated by time τ, and $m(t_j)$ is the magnitude at epoch t_j. The power density spectrum is arguably the most powerful tool for studying variable light curves.

Figure 3.6 is a $\log_{10}$-$\log_{10}$ plot of P(s) against the variability as seen by observation from Earth. Increasing values along the horizontal axis represent a shorter period of variation. The longest periods for variability are found at the highest power levels, as expected. Sample data for the power spectrum analysis came from two different surveys. The first source of data was the ESO/SRC 287 survey from which 810 samples (filled circles) were chosen. The second source was the massive compact halo object (MACHO) project where 68 quasars (open circles) were chosen from within the required luminosity bounds. The variations in flux are be-

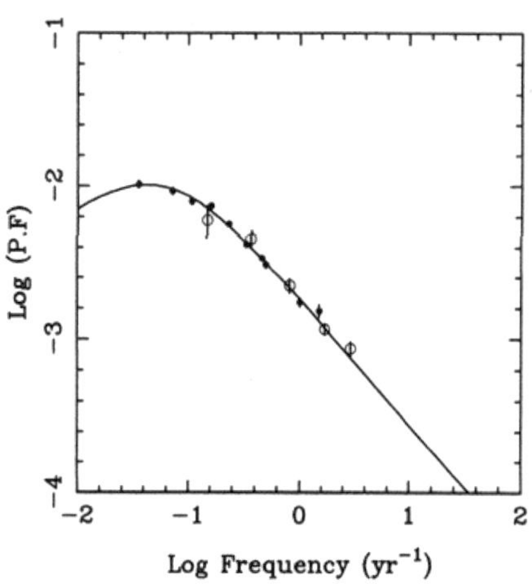

Figure 3.6

* Cosmological time dilation provides evidence for an expanding universe.

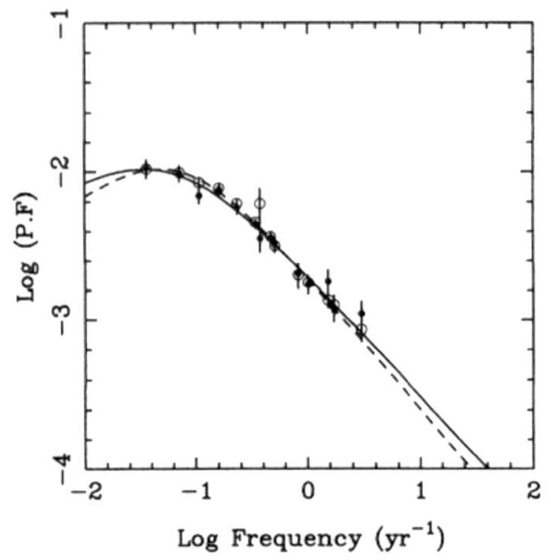

Figure 3.7

lieved to be intrinsic to the quasar. As such, they are subject to cosmological time dilation.

A consequence of general relativity is that any wavelength of light that has shifted by a factor of z will be observed with a time dilation factor of $(1 + z)$. If the data in the previous graph were separated into high- and low-redshift quasars, we would expect divergence in the frequency scale factor of the two newly formed plots. Figure 3.7 contains the plots of quasars with absolute magnitude restriction $-25.5 < M_B < 22.5$ at $z < 1.0$ (solid line) and $z > 1.0$ (dotted line). The two plots are tightly correlated, and we do not see the expected shift of the high redshift sample towards longer time scales relative to the low redshift sample. This stands in stark contradiction to the expected effects of general relativity.

Various theories have been offered (and failed) to explain this unexpected phenomenon. First is that black holes grew in size over time and the quasars thereby decreased in variability. The time dilation of the distant ($z \gg 0$), smaller quasars would get offset exactly by the closer ($z \sim 0$), larger quasars. The narrow absolute magnitude of the sample, however, ensures a small variance in the black hole mass. A second explanation is that the variations are not predominantly intrinsic to the quasars. Microlensing by stellar mass bodies, for example, could have an effect. This explanation fails for lack of an observed population of objects along the line of sight. Of course, there is always the explanation that there is some unknown selection bias in the data.

It is possible that the sample somehow contains both closer quasars with longer periods and more distant ones with shorter periods. By 'lucky coincidence' their observed periods closely match and negate the effects of time dilation. This explanation also fails upon closer scrutiny. Figure 3.8 shows the quasar luminosity plots trisected by high (filled dots), medium (open dots) and low (asterisks) bins. The 'asterisk' curve has the greatest range of brightness variation as expected with the less luminous quasars. The plots from these luminosity bins have a measurable difference in slope not found in the plots of Figure 3.7 The 'lucky coincidence' is ruled out. Rourke (2014) thus deems that "these results conclusively prove that intrinsic redshift exists in quasars." [64]

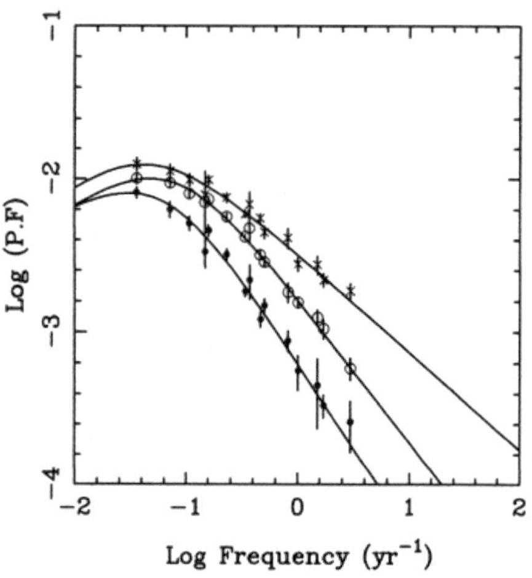

Figure 3.8
© 2010 Oxford University Press

What if there did exist an intrinsic component to the redshift? Perhaps portions of the accretion disk lie at varying gravitational strengths. What is measured as greater luminosity may actually be some measure of greater strength inside the gravitational well. The difficulty with this explanation is that it goes against the widely held belief that redshift represents distance, and is strictly cosmological in origin.

Age Crisis

A recurring problem in the field of cosmology has been the discovery of astronomical objects that are dated older than the established age of the universe. An object at redshift z is constrained in age to less than the epoch age, $EA = t_0/(1 + z)^{3/2}$, where t_0 is the current age of the universe. For example, the discovery of a galaxy at $z = 3$ with a stellar population of 2 billion years would require $t_0 > 2(1 + 3)^{3/2} = 16$ billion years. Of course this would contradict the *known* age of the universe. The cosmological constant, Hubble constant, matter density, cold dark matter, dark energy, *etc.*, all tuned to fit the most popular cosmological model, ΛCDM, yield an age of 13.8 billion years for the universe. The oldest known local objects are the globular clusters whose age has been calculated at 13.4 billion years. [65]

The discovery of active galactic nuclei that pre-date the Big Bang reintroduces the age crisis. APM 08279+5255 at $z = 3.91$ (age 2-3 Gyr), LBDS 53W069 at $z = 1.43$ (age 4 Gyr), and LBDS 53W091 at $z = 1.55$ age (3.5 Gyr) are some examples whose measured values yield an existence *before* the

Big Bang. Komossa and Hasinger (2002),[66] Hamann and Ferland (1993),[67] and Friaca, Alcaniz and Lima (2005)[68] have analyzed the age of APM 08279+5255 using X-ray observations, the giant elliptical model (M4a), and a chemodynamical model for the evolution of spheroids, respectively. All conclude that the quasar has a minimum age of 2 Gyr. That makes it older than the Big Bang by nearly one billion years.

Given the calculated age and distance of APM 08279+5255, it became necessary for researchers to study the impact of modifying the parameters of ΛCDM. Deepak Jain and Abha Dev[69] published the categorical statement: "It is alarming to note that the quasar cannot be accommodated in any dark energy model even for $\Omega_m = 0.23$, which corresponds to 1σ deviation below the best fit value provided by WMAP." Some theorists have used this powerful evidence to advance new cosmologies. A simpler approach is to assume an intrinsic redshift of the quasar.

Superluminal Motion

It has been known for 40 years that there are radio sources with small-scale components that appear to separate from each other with a velocity that exceeds the speed of light. The earliest evidence for such motion came in 1971 when the double structure of radio galaxy 3C 273 was observed to separate with a linear velocity 10 times the speed of light over a period of a few months. Since that time, HST has imaged the object repeatedly. 3C 273 owes its fame to the fact that it was the first quasar ever discovered. It also

The Illusion of Superluminal Motion

A cloud of ejecta begins its journey at time $t = 0$ with relativistic velocity v along a trajectory at an angle θ to the line of sight of an observer.

After time t_2, the ejecta has traveled a distance vt_2 along its trajectory.

Light from the initial ejection reaches a distance

$$ct_1 = vt_2 \cos\theta$$

at time t_1 and a distance ct_2 at time t_2.

From time t_1 to time t_2, the ejecta has appeared to move transversely by the distance

$$dx = vt_2 \sin\theta .$$

The time difference $dt = (t_2 - t_1)$ can be related to the distance traveled by the light from the initial ejection

$$c(t_2 - t_1) = ct_2 - vt_2 \cos\theta$$

and thus

$$dt = (t_2 - t_1) = t_2 - \frac{vt_2 \cos\theta}{c} .$$

The apparent transverse velocity of the ejecta as seen by an observer is then

$$v_{app} = \frac{dx}{dt} = \frac{vt_2 \sin\theta}{t_2 - \frac{vt_2 \cos\theta}{c}} = \frac{v \sin\theta}{1 - \frac{v}{c}\cos\theta} .$$

For large v and small θ, v_{app} will exceed the speed of light.

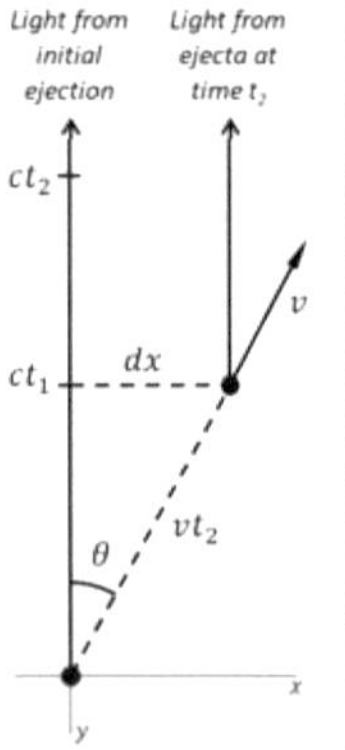

Figure 3.9 - Geometry of superluminal motion

holds the more problematic position as the centerpiece in the debate on superluminal motion.

Although the special theory of relativity tells us that quasar components cannot actually travel faster than the speed of light, Stuart Clark has offered an explanation how it appears to happen.[70] Figure 3.9 explains the illusion in mathematical terms. The time interval between two beams of light as measured by an observer on Earth is not equal to the actual interval between emissions by the jet moving at relativistic velocity. The first beam has had a longer path to travel. We are thus in a position that underestimates the interval of time between the two emissions, and we will measure a velocity that is *apparently* faster.

It is customary to use the dimensionless variable $\beta = v/c$ (where c is the speed of light) when representing the real transverse velocity of the relativistic object. Upon substitution,

$$v_{\text{tran}}/c = \beta_{\text{app}} = \beta \sin\theta / (1 - \beta\cos\theta).$$

The sketch on the right reveals a heavy dependence on large β and small θ for superluminal motion, β_{app}. By statistical argument only 2% of the studied radio emitting quasars should be aligned in such a manner.[70] This stands in contrast to the fact that the majority of such objects show sizable β_{app}. There is a possible explanation for this discrepancy.

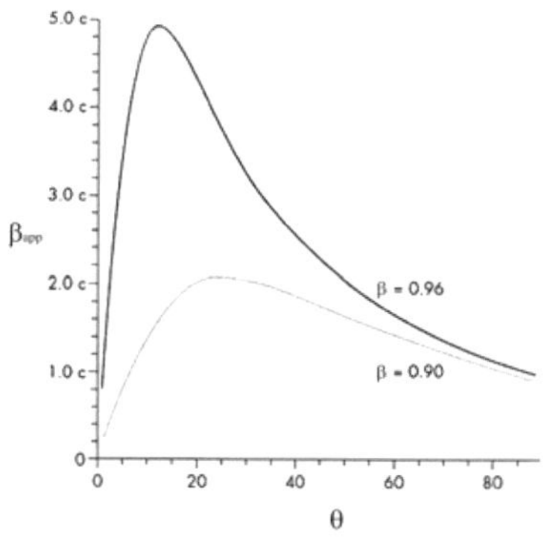

The radio-quiet quasars outnumber the radio-loud by 100 to 1. If radio emission is detected only from those quasars whose beams are tilted in the direction of Earth, then the high percentage of superluminal observations is understandable. This explanation, however, does not account for observations of superluminal jets parallel to the plane of the sky.

Narlikar and Chitre (1984) were early critics of the "line-of-sight beaming" explanation of superluminal motion of quasars.[71] They argued that the predicted intensity enhancement is not seen for the near "blob" of a dual jet quasar. The beam model predicts a near-blob intensity gain by a factor

$$f = \left[\gamma\left(1 - \beta\cos\theta\right)\right]^{-(3+\alpha)},$$

due to the Doppler effect, where γ is the Lorentz factor and α is the spectral index of the radiating source. If both blobs were comparable in luminosity in their respective rest frames, the near blob would appear ~10 times brighter than the far blob. Upon actual observation the brightness of the two blobs is not often significantly different. For 3C 273 they calculate the low probability of $P \sim 10^{-5}$ that beaming is the actual cause of the *apparent* superluminal motion. Richard Porcas (1983) presented another problem when he revealed that the average outer structure of double-lobed superluminal radio quasars is not smaller than that of the normal radio-source population. [72] This then implies that the outer structure is not at a small angle to the line of sight of the inner jet unless there is misalignment between jets. Most troubling of all were the findings of Thomson, Mackay and Wright (1993). [73] They superimposed the optical jet of 3C 273 onto a contour plot of a reconstructed HST image to show the orientation of the magnetic-field vectors. For optical emissions from 3C 273 scattered through a jet at small angle to the line of sight, the magnetic polarization vectors should be everywhere parallel to the radius vector to the quasar. However, the data show significant deviations from a parallel orientation. They also showed that the optical radiation losses become significant at the point where the jet interacts with an edge-brightened shell of gaseous material. This further implies the jet cannot be aligned closely to the line of sight. They came to the alarming conclusion that "3C 273 must be viewed in the plane of the sky nearly perpendicular to the line of sight."

In a more recent publication (Lister *et al.*, 2009) [74] the jet kinematics are studied for a flux-density-limited sample of 135 radio-loud AGNs taken from the MOJAVE survey. The sample was selected on the basis of compact (2 cm VLBA) flux density, which favors AGNs that have strongly beamed jets viewed at small angles to the line of sight. Histogram 3.1 is a distribution of 502 robust jet features. The shaded region represents the distribution of the fastest robust feature in each jet. The fastest identified jet component (in quasar 0805-077) measures $50.6c \pm 2.1c$. This is solid support for the beam model. When the line-of-sight angle is small toward a jet moving close to the speed of light, we will see *apparent* motion greater than *c*.

Although the great majority of the sources in the MOJAVE sample have one-sided jets (not unexpected given the large Doppler boosting), a few sources show two-sided jets which are considered to lie close to the plane of the sky where there is negligible Doppler boosting. The beam model necessitates an apparent speed of $\beta < 1$ for these objects, but this is not always the case. Superluminal jet features are found in 1413+135 (~1.8*c*), 3C 120 (~4*c*), M87 (~6*c*) and 3C 273 (~15*c*). Given the maximum speeds documented for the entire MOJAVE sample of VLBI-observed jets, the maximum observed speed of 15*c* of the 3C 273 parsec-scale jet implies an absolute maximum angle no larger than 7.2° (assuming $\beta = 1$) to the line of sight. This stands in stark contrast to all the evidence that 3C 273 lies in the plane of the sky nearly perpendicular to our line of sight.

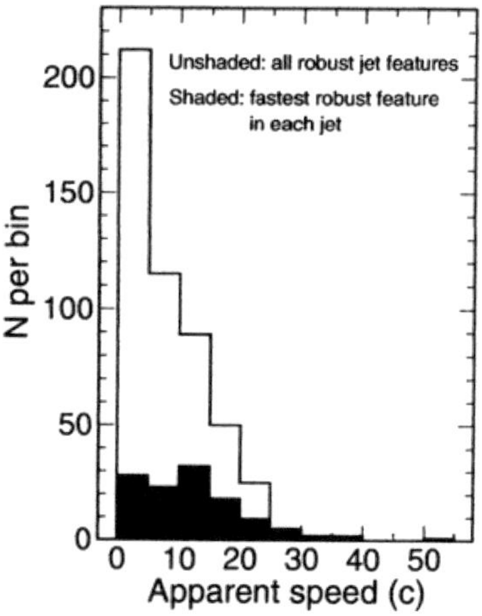

Histogram 3.1 – Robust jet features

Explanation is lacking also for 3C 273 X-rays. Recent *Chandra* observations show that the X-ray jets are inconsistent with the known radio-optical synchrotron spectrum. The model to explain the anomaly is that the X-rays are produced by inverse Compton up-scattering of CMB photons by a jet highly relativistic at kpc scales. The jet should match the high speeds of the knots in jets like 3C 273. However, this model has recently been ruled out based on gamma-ray upper limits and strong upper limits on knot proper motion revealed by Meyer *et al.* (2016).[75]

In their later MOJAVE paper Lister *et al.* (2013) analyzed other factors contributing to the superluminal motion.[76] They point out that the "spread in apparent speed we see in individual jets cannot be wholly attributed to a spread in streamline viewing angle." Figure 3.10 plots apparent speed *vs.* luminosity for various groups of objects. It is possible that the positive correlation is related somehow to greater jet power. It also is interesting to consider that an overestimated distance calculation would boost both the apparent speed and luminosity.

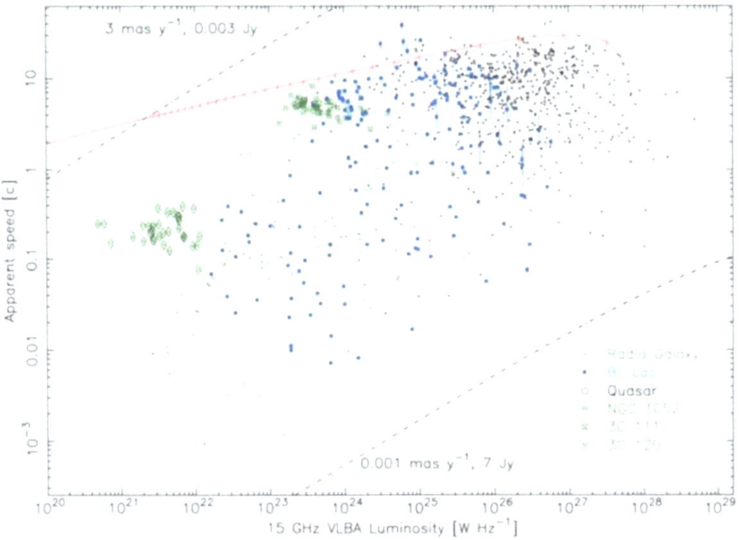

Figure 3.10 – Apparent speed vs. luminosity

Preferred Redshift and Periodicity

The early belief in an isotropic universe held that extragalactic objects are found at random distances one from the other. This was soon tempered by the realization that galaxies tend to cluster within groups. When objects now are found separated one from the other at equal length, they are said to have a *preferred* distance interval. When objects now are found separated one from the other at some quantum multiple of a fixed length, they are said to have distance interval *periodicity*. When distances are determined by their redshift, these intervals are known respectively as *preferred redshift* and *redshift periodicity* (or *quantization*). The subject of redshift periodicity evokes much confusion. One reason is that the implied source of the measurement is not always the same. The distances of nearby galaxies from the center of the Milky Way determine the possible *galaxy periodicity*; the distances of superclusters far away from Earth determine the possible *supercluster periodicity*; the distances of quasars from their respective 'assumed' parent galaxy determine the possible *quasar periodicity*. Each category of redshift periodicity has brought its fair share of debate. The latter was most contentious given its potential to disrupt the cosmological hypothesis.

William G. Tifft is well known as the pioneer who revealed the strange phenomena associated with redshift for 108 galaxies in the Coma cluster. He began his work at Steward Observatory in 1970, studying redshift correlations to galaxy luminosity, morphology, and activity (radio and optical emission).[77] As measurements improved he discovered that the redshift values had a periodic nature. *Redshift Quantization* was formally presented in a series of three papers in the *Astrophysical Journal* (1976-1977).[78] [79] [80] The second paper (Tifft, 1977a)[79] announced that redshift differentials between galaxies in clusters contain "preferred values which are various multiples of a basic 72.5 km s^{-1}." This observation is at odds with conventional dynamics, which predicts a smooth monotonic distribution. Further studies strongly confirmed the original prediction.[81] [82] [83] [84] [85]

Interest then turned toward a possible globally linked pattern in redshift correlation. This would require reliable redshift data across different parts of the sky, a complicated matter due to the Doppler component added by the Sun's motion. The data must be accurate to within 10 km s^{-1} in order to properly detect the presence of any real periodicity. This stringent requirement is met by 21 cm radio observations of H_I clouds. Tifft and Cocke (1984) used the Fisher-Tully 21 cm data to show that 1000 galaxies are strongly periodic at 36.2 km s^{-1}.[86] Croasdale (1989) used 157 wide profile galaxies from the Arecibo Observatory 21 cm data to support the hypothesis that galaxy redshifts occur in steps of slightly more than 72 km s^{-1} or a simple sub-multiple thereof.[87]

Further support of galaxy periodicity came from two astronomers who initially set out to disprove the Tifft hypothesis. Napier and Guthrie conducted a series of tests with high precision H_I profiles of nearby bright galaxies. They consistently found the redshift distribution to be strongly periodic in the galactocentric frame of reference. The first study used galaxies within 10° of M87 near the core of the Virgo cluster.[88] They were divided into two samples: 112 spirals and 77 dwarfs with redshift < 3000 km s^{-1}. The sub-sample of low-density spirals as a function of redshift is shown in Figure 3.11.

The strong periodicity is easily seen. The confidence level is at $0.996 < C < 0.999$ on the assumption that the Local Group is falling directly towards the Virgo cluster. The phenomenon cannot be ascribed to statisti-

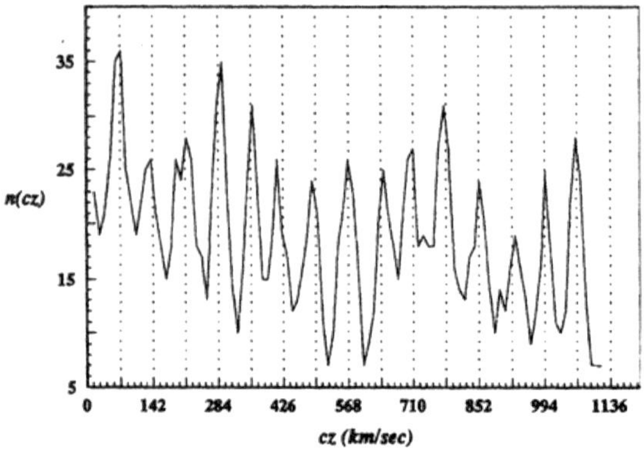

Figure 3.11 – Galactocentric redshifts of low-density Virgo spirals

cal artifacts, selection procedures, or flawed reduction techniques. Subsamples of the high-density spirals and dwarf galaxies did not show any periodicity.

The second study made use of the galaxies in the Bottinelli catalogue. Forty bright spirals with redshift $< 1000\ \mathrm{km\ s^{-1}}$ revealed a periodicity in the range 37.2-37.5 km s^{-1}. [89] Again no evidence was found for periodicity in the irregular galaxies. The conclusion is that nearby field galaxies have a redshift periodicity of ~37.5 km s^{-1} relative to the center of the Galaxy. The third study extended the enquiry to the edge of the Local Supercluster. [90] It found the ~37.5 km s^{-1} periodicity is maintained out to the edge at 2600 km s^{-1}. Figure 3.12 shows the periodicity as far out as 1500 km s^{-1}. These studies have never been disproved. Perhaps it is the case that some unknown physics causes the galaxy periodicity, at least within the Local Supercluster. It is not exactly clear why more effort has not been made to explain it. While it strains the cosmological hypothesis, it does not dis-

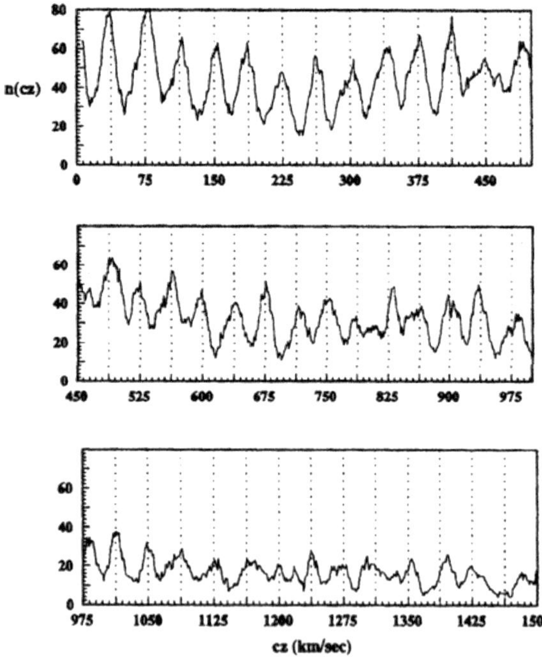

Figure 3.12 – Galactocentric redshifts throughout the Local Supercluster

prove it. Analysis of supercluster periodicity and quasar periodicity may help provide an answer.

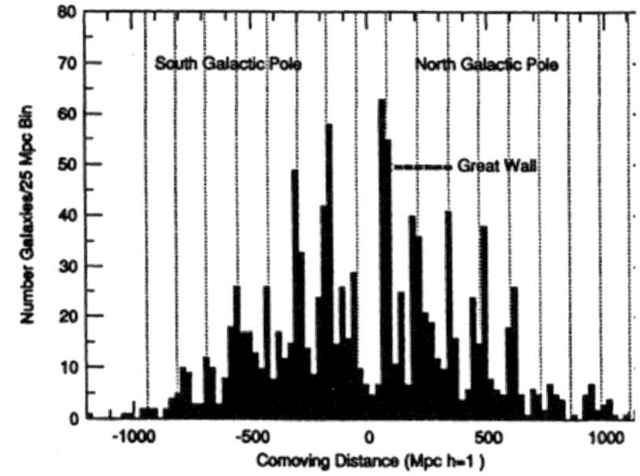

Figure 3.13 - Supercluster periodicity

Low-depth all-sky surveys led to the discovery of galaxy periodicity. Narrowly focused deep space surveys, known as pencil beams, revealed the universal periodicity of the superclusters. The pencil beam surveys small regions of the sky deep into the universe to an extremely faint optical limit. The first four distinct surveys were restricted to a few arcminutes from the north and south galactic poles. Koo *et al.* (1993)[91] combined their data with Broadhurst *et al.* (1990)[92] to show regular maxima in the redshift distribution with a characteristic scale of 128 h^{-1} Mpc extending to 2000 h^{-1} Mpc (where $h = H_0/100$ km s^{-1} Mpc^{-1}, and H_0 is the Hubble constant). As seen in Figure 3.13, the smooth distribution assumed by the cosmological hypothesis does not occur with increasing distance.

Later analysis (Saar *et al.*, 2002) found that the supercluster-void network resembles a cubic lattice of great regularity out to the distance of 350 h^{-1} Mpc.[93] The observed supercluster periodicity presents another difficulty for the standard cosmological model. The recurring 128 h^{-1} Mpc distance between walls in every direction from Earth belies the cosmological principle of an isotropic universe. Some theorists believe that the periodicity of nearby galaxies may be some form of sub-structure of the supercluster-void cosmic sponge. Any solution will require more than conventional dynamics (with or without dark matter), and that will further weaken CH. If quasar redshift periodicity is shown to exist, then the CH anvil will crack.

The quest to demonstrate the existence of preferred redshift or periodicity among the quasars has been a formidable task. Few people remain who believe it necessary to do so. They simply accept that quasars exist at their cosmological distances. Yet the accumulated observations suggest that something important is being ignored. Geoffrey and Margaret Burbidge were early advocates of some possible intrinsic gravitational redshift in the objects now known as quasars. In their book, *Quasi-Stellar Objects,* [94] they claimed that many QSOs have a standard absorbing region with a preferred redshift of 1.95. Soon other preferred redshifts were discovered. The Karlsson quasar redshift distribution followed. Karlsson (1977) revealed a highly significant period of $\log(1+z)$ in the distribution of 574 quasar emission line redshifts, and further showed that this is not due to observational selection. [95] His formula had the form:

$$(1+z_2) = 1.23 \cdot (1+z_1),$$

which yields $z_n = 0.061, 0.30, 0.60, 0.96, 1.40, 1.96\ldots$

These are the famous Karlsson redshifts. Fang *et al.* (1982) confirmed the distribution at 99% confidence using an updated catalogue of 1429 quasars. [96] If these data were a true representation of the cosmos, it would suggest something extraordinary.

One of the early objections to the Karlsson and Fang data was that it contained insufficient sampling. Some researchers did modify the Karlsson formula as more complete samples of data became available, but some form of periodicity always remained.* Later studies also began to transform the redshifts to the galactocentric frame. Using the compilation (Hewitt & Burbidge, 1990) of 2164 QSO redshifts, Duari *et al.* (1992) [97] rejected the Karlsson formula in favor of a simple periodicity of $\xi = 0.0565$, a value surprisingly close to that offered by Burbidge twenty-four years earlier. [98] Using a sub-sample of approximately 5000 high-redshift quasars from the SDSS Quasar Catalog, Bell (2004) proclaimed the Karlsson formula invalid for $z > 2$ and introduced the *decreasing intrinsic redshift model* (DIR model). [99] DIR builds on the more radical theory that quasars are ejected from low-z galaxies and that they have an *intrinsic* redshift component that is quasi-periodic given by the formula

$$z_{iQ} = z_f\left[N - M_n\right],$$

where N is an integer, M_n is a function of a secondary quantum number, and $z_f = 0.62 \pm 0.01$ is the intrinsic redshift constant. The predicted intrinsic redshifts are shown in Table 3.3 (for $z < 4.5$). Notice that the first three

* Recent observations favor $(1+z_2) = 1.228\ (1+z_1)$, which yields $z_n = 0.060, 0.302, 0.598, 0.963, 1.410, 1.960,\ldots$

(n)	Z_{iQ}[N = 1,n]	Z_{iQ}[N = 2,n]	Z_{iQ}[N = 3,n]	Z_{iQ}[N = 4,n]	Z_{iQ}[N = 5,n]	Z_{iQ}[N = 6,n]	Z_{iQ}[N = 7,n]
0	0.620	1.240	1.860	2.48	3.10	3.72	4.340
1	0.558	1.178	1.798	2.418	3.038	3.658	4.278
2	0.496	1.054	1.488	1.178			
3	0.434	0.868	0.558				
4	0.372	0.620					
5	0.310	0.310					
6	0.248						
7	0.186						
8	0.124						
9	0.062						

Table 3.3 – Intrinsic redshifts predicted by DIR model

Karlsson numbers are approximated by combining $N = 1$ with $n = 9$, 5 and 0, respectively.

Since the DIR model contains both the *preferred* intrinsic components and the *smooth* Hubble flow components, all peaks are predicted to show an asymmetry toward the red wing. The bold histogram in Figure 3.14 gives the distribution of the ~5000 quasars under study by Bell (2004). For ease of view the dashed curves indicate peak profiles that reproduce the histogram. The dotted curves separate cosmological components from the ejection components whereby the extended red wing asymmetry is seen more easily. The vertical dashed lines indicate the positions of the predicted preferred redshifts. Such a density clumping is difficult to explain in a strictly CH universe, but the proposition that low-*z* galaxies are parenting high-*z* quasars does not meet with much approval. Larger quasar samples were necessary to resolve the issue.

A long list of research papers appeared in the 20th century, but the issue of quasar periodicity was not satisfactorily resolved. The difficulty of quasar detection had resulted in samples that were undersize and subject to possible selection effects. The 21st century promised a treasure trove of data from two new large-scale surveys, 2dF and SDSS, that might improve the detection.

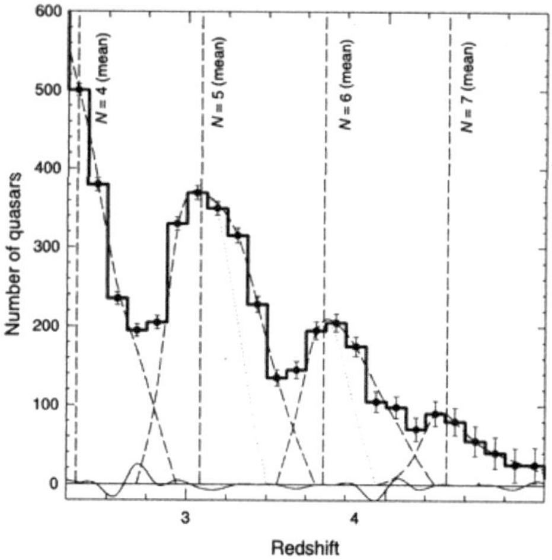

Figure 3.14 - Redshift distribution
(2.4 < z < 4.8)

After a successful confirmation of the redshift periodicity in new samples of QSO pairs (or QSO multiples) found close to foreground galaxies,[100] William Napier approached Edward Hawkins for an unbiased analysis on the much larger homogeneous sample of data from the 2dF Galaxy Redshift Survey (2dFGRS) and the 2dF QSO Redshift Survey (2QZ). Hawkins considered only those galaxies in the range $0.01 < z < 0.3$ and only those QSOs with the highest quality flag in the database. This left 67,291 galaxies and 10,410 QSOs, from which 1,647 QSO-galaxy pairs were found. After rigorous statistical analysis, he found no evidence for a periodicity at the frequency predicted by Napier *or any other frequency*. Hawkins *et al.* (2002) concluded "Given that there are almost eight times as many data points in this sample as in the previous analysis by Burbidge and Napier (2001), we must conclude that the previous detection of a periodic signal arose from the combination of noise and the effects of the window function*."[101] A response came quickly. Napier and Burbidge (2003) described a methodology that could assess these selection effects and found that the failure to detect the periodicity in the 2dF data must have been due to some significant difference between the data sets.

> One obvious difference is that the galaxies in the Burbidge-Napier (BN) sample are bright, nearby, active spirals, whereas those in the 2dF survey are an inhomogeneous† selection going out to $z \sim 0.3$. In a survey such as the 2dF, a major problem is that the galaxies which are assumed to be paired to the QSOs have a wide range of cosmological redshifts, perhaps up to $z_c \sim 0.3$. On a local ejection hypothesis, the nearest galaxy to a QSO may not be the parent galaxy, so the correct cosmological redshift term to be removed may not be known.[102]

In other words, all that extra 2dF data help only if we learn how to carefully select parent galaxies lest the periodicity peaks become smeared out by cosmological components!

There is no significant difference between the data set used by Karlsson (1990) and that adopted by Burbidge and Napier (2001). Histogram 3.2 shows the combined data sets. The periodicity is quite evident, with all peaks at Karlsson numbers. The original Karlsson periodicity cannot be

* The 'window function' truncates an idealized infinite data series to a finite series. If the cut-off occurs sharply, strong spurious features can be introduced.

† Inasmuch as bright spirals were the only galaxy types for which *galaxy periodicity* had been found, it is curious that Napier did not instruct Hawkins to limit the galaxy data set accordingly.

ascribed to any observational artifact and was found to be present at the very high significance level of approximately 10^{-5}; The Burbidge-Napier data cannot be ascribed to statistical artifact. While the evidence for the combined 290 QSO redshifts is rather convincing, it has not been determined whether the effect is universal or caused by some other unknown artifact.

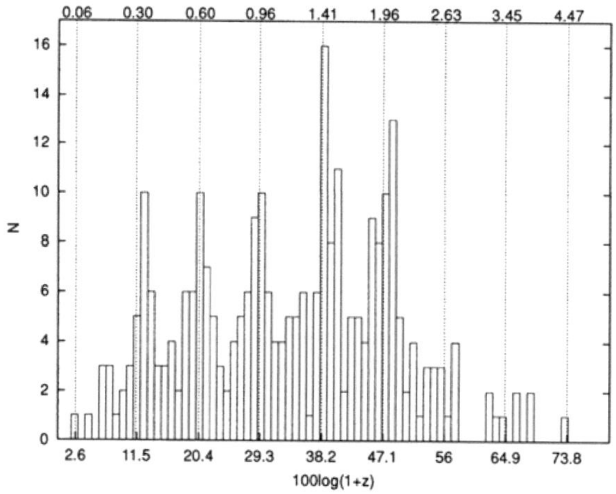

Histogram 3.2 - QSO data from Karlsson and B-N

A similar situation ensued when data from the SDSS survey were added to 2dF for analysis. The most famous attempt to disprove QSO redshift periodicities came from Tang and Zhang (2005).[103] They added the 46,400 QSOs found in SDSS Data Release 3 (DR3) to those in 2QZ to show that QSOs are not likely to be ejected by nearby active galaxies, and found no evidence for any periodicity in their redshifts. The quick response by Bell and McDiarmid claimed that Tang and Zhang probably missed, or misidentified, many of the parent galaxies.

> Tang & Zhang (2005) did not detect a significant power peak near $\Delta z = 0.62$ (because) they did not correct for the effects of strong low-frequency components when they examined the lower half of the redshift data…..and because they extended the redshift range down to $z = 2$, which covers a portion of the much more highly populated low-redshift source sample, in the high-redshift sample…..That there is little evidence for a period in the 2QZ redshift distribution can be explained if that survey found more sources with larger cosmological components present in their measured redshift, filling in the valleys and smearing out the peaks…..[104]

Would more data help end the debate? Repin *et al.* (2012) analyzed the enlarged database of SDSS DR7 and further asserted an absence of any periodic component in quasar distribution.

> We conclude from the analysis that we have not been able to reliable (*sic*) identify a regular periodic component in the quasar redshift distri-

> bution. However, the rippling of the quasar density probably exceeds the statistical uncertainties. It is most likely that we are dealing with a cellular structure. The individual cell walls appearing in the line of sight contribute to the number of quasars at the corresponding z values. Further supplementing of the (SDSS) catalog should smooth irregularities, making it possible to distinguish quasi-periodicity in the quasar z distribution at increasing significant confidence levels. [105]

But this report also received quick criticism. No periodicity signal will be found unless there is a proper method to determine the parent galaxies. With well defined constraints on each candidate quasar, Fulton *et al.* (2018) found a strong signal for redshift periodicity in the data from SDSS DR7 (and 2MRS). Their conclusions stand in stark contrast to Hawkins *et al.* (2002), Tang and Zhang (2005), and Repin *et al.* (2012).

> …certain SDSS quasars are physically associated with lower redshift SDSS galaxies and separately with lower redshift 2MRS galaxies; at least some quasars of very different redshift are physically associated with the same nearby galaxy; ….. A quasar excess exists at Karlsson redshifts around the 2dF, SDSS, and 2MRS galaxies. [106]

This is the latest report on the matter. It has not been repudiated, but it also has gained little recognition. Nobody denies that there exist unexplained phenomena surrounding quasars, but redshift periodicity is typically excluded from the list. Even the periodic redshift distribution of the Local Supercluster galaxies is not universally accepted. While *periodic* redshifts may be dismissed, the problem of *preferred* redshifts remains. The value $z = 0.62$ is found consistently, even by those who find no pattern in the overall redshift distribution. It is difficult for CH to assume that quasars formed more frequently at a certain epoch.

Near Neighbor - Revisited

The best argument against CH is that high-z quasars are found in proximity to low-z galaxies. It is then very odd that CH advocates claim the exact opposite in support of their position. The fact is that both sides have statistical studies that support their respective claims. The CH detractors, however, claim additional evidence by way of images that show observable links between high-z quasars and their low-z nearby neighbors.

Halton Arp (2001) published the opinion: "It has been well established that quasar redshifts occur in certain preferred peaks. These peaks are given by the Karlsson formula." [107] The historical record does not sup-

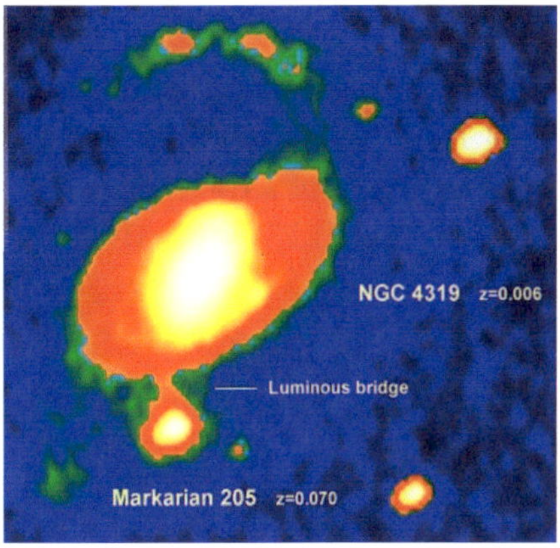

Figure 3.15 – NGC 4319 and Markarian 205

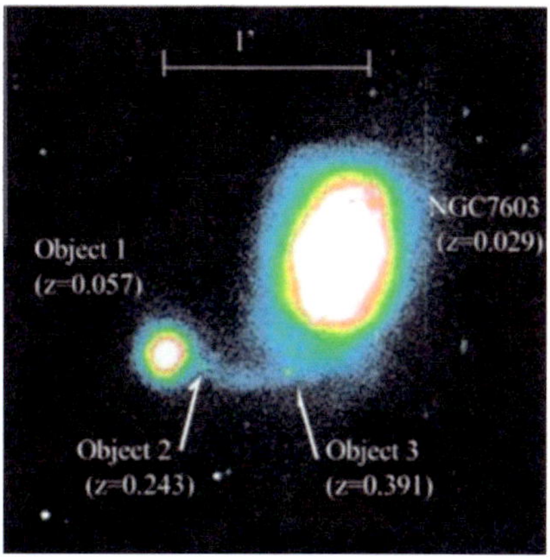

Figure 3.16 – NGC 7603
© 2002 EDP Sciences

port such a strong claim; many skeptics remained steadfastly against any theory of discordant redshifts. No doubt Arp felt emboldened by his large collection of photographic evidence linking quasars to nearby galaxies.

For example, his 1998 book *Seeing Red*[108] exposes numerous images of galaxy-quasar associations that stand in refutation of cosmological redshift. The most famous example is the connected pair of objects NGC 4319 ($z = 0.006$) and Markarian 205 ($z = 0.070$). Figure 3.15 reveals the bright physical bridge between the objects. Professional astronomers dismissed the *connection* by claiming it is caused either by some intervening object or by overlapping isophotes between the images of NGC 4319 and Markarian 205. Sulentic (1983) ruled out these explanations and further showed that the connection can be traced into the nuclear regions of NGC 4319.[109] Most other Arp examples were not as clear when first published, but later research provided excellent confirmation. Lopez-Corredoira and Gutierrez (2002), for example, offered a clear image of the optical filament connecting high-z objects to NGC 7603,[110] as seen clearly in Figure 3.16.*

Figure 3.17 – NGC 7319

* Chapter 5, **Luminous Bridges**, discusses the more recent study of the luminous NGC 7603 bridge

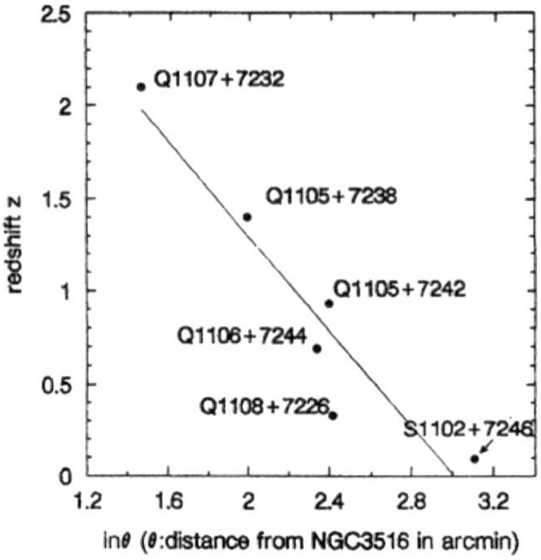

Figure 3.18 – z vs. θ for NGC 3516 group

A second type of image that refutes CH is a high-z quasar seen *forward* of the low-z galaxy. Figure 3.17 shows the image of a quasar ($z = 2.11$) in front of NGC 7319 ($z = 0.022$).[111] There are no noise factors that will explain the foreground quasar. A third type of image shows multiple quasars aligned to the axis of the parent galaxy. Chu *et al.* (1998) showed the alignment of five quasars along the minor axis of NGC 3516 ($z = 0.09$).[112] More significant is the fact that the quasar redshifts correlate to angular distance, where the highest redshift is closest to the parent NGC 3516. Figure 3.18 shows the correlation. Arp characterized the NGC 3516 group as the *Rosetta Stone of intrinsic redshift*.

It is difficult to refute the powerful images, but they represent a very small sampling of quasars in the Universe. Those who support the cosmological interpretation of quasar redshifts remain firm in their belief that some yet undetermined factor(s) will explain the observed anomalies. Arp was equally strong in his belief that the evidence supported discordant redshifts. His persistence did not sit well with the CH camp, who claimed that Arp relied upon *a posteriori* statistical analysis and had no coherent theory to hold together all of his explanations for the peculiar galaxies. Instead of accolades for his discoveries, Arp was roundly criticized and cut off from further viewing time at the Palomar Observatory. Daniel Weedman went further by stating, "I do not believe the prospect of finding non-cosmological redshifts to be sufficiently promising that more effort should be invested in the task."[113] If Arp's gallery of pictures represented too small of a sample to form any conclusive opinion, and if further observations were restricted,* an updated statistical analysis was necessary.

The work of Geoffrey Burbidge is unparalleled in the study of QSO-galaxy close pairs. Before the discovery of QSOs, Burbidge already had earned high respect from astrophysicists for his collaboration on the fun-

* Burbidge *et al.* (2004) did provide further observations.[115]

Geoffrey Burbidge (1925-2010)

damental paper on stellar nucleosynthesis, "Synthesis of the Elements in Stars." [114] Once discovered, QSOs held his attention until his death in 2010. When the early work of Arp suggested that QSO redshifts were not of cosmological origin, Burbidge decided upon a clean statistical test. Using 6σ significant samples from the 3C radio catalog and bright galaxies in the Shapley Ames catalog, he discovered that the spatial distribution of the 47 identified QSOs was much closer to the small-*z* galaxies than would occur if they were distributed randomly. Burbidge *et al.* (1971) showed that such a chance occurrence has a probability less than 5×10^{-3}. [116] They also showed that the angular separation θ between the QSO and (nearby) galaxy is proportional to the redshift of the galaxy.

If there truly is a redshift discrepancy in QSO-galaxy pairs, perhaps it exists for QSO-QSO pairs as well. Discordant QSO-QSO pairs would lend support to advocates of both non-cosmological redshift *and* the quasar redshift periodicity. By 1985 there was a total of 30 documented QSO-QSO pairs with angular separation of 120 arc-seconds or less. Burbidge *et al.* (1985) documented 6 of these pairs with near identical redshifts (believed to be gravitationally lensed) and the remaining 24 pairs with discordant redshifts. The probability for such a chance occurrence is again very low at $P < 10^{-4}$. [117] Burbidge's findings of redshift discrepancy in both QSO-galaxy pairs and QSO-QSO pairs did not find favor with those who had recently made a formal declaration for cosmological redshift.* A larger sampling of data would be necessary.

By the end of the decade the number of known QSOs had grown to ~3000. Burbidge, Hewitt, Narlikar and Das Gupta embarked upon another statistical analysis of QSO-galaxy pairs using galaxies from *The Revised New Catalog of Nonstellar Astronomical Objects.* [118] The published result (Burbidge *et al.* 1990) of the updated analysis triggered another quake in the astronomical community. It contains a table of all close pairs of 577

* See introduction to Chapter 2.

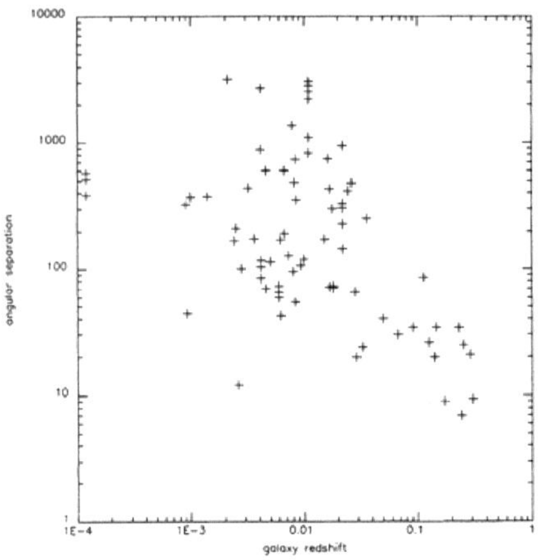

Figure 3.19 – θ vs. z_G
(76 QSO-galaxy pairs)

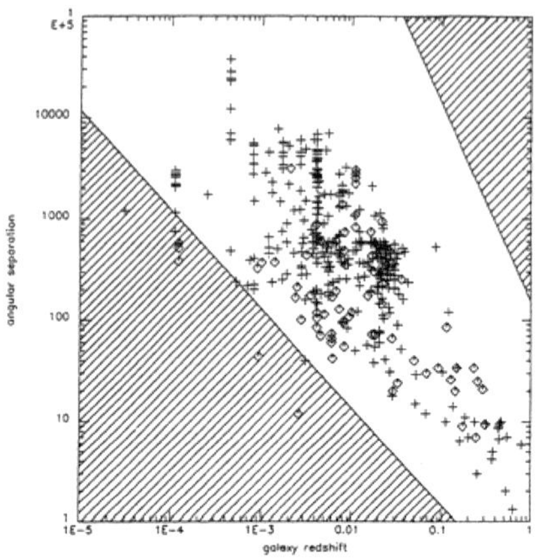

Figure 3.20 – θ vs. z_G
(392 QSO-galaxy pairs)

QSOs and more than 500 galaxies for which the majority of the QSO-galaxy angular separations is less than 600 arc-seconds.[119] The table contains 28 QSOs associated with 42 galaxies at the same low-*z* redshift. All the other pairs are at $z_Q \gg z_G$. This provides *prima facie* evidence that high-*z* QSOs are often *near neighbors* of low-*z* galaxies. Their paper further substantiates the proportionality between the angular separation θ and the galaxy redshift z_G. Figure 3.19 shows the observed relationship in 1980 for Burbidge's 76 QSO-galaxy pairs. Figure 3.20 shows the updated relationship with 392 QSO-galaxy pairs. A projection effect causes some of the observed scatter. The hatched areas are regions excluded by selection effects. The correlation is strong evidence that the majority of pairs are physically associated and that QSOs often have a dominant intrinsic redshift factor. The paper concludes: "QSOs tend to lie in the vicinity of normal galaxies much more often than is expected by chance, *whether or not* the galaxies and the QSOs have the same redshift."

Analysis did not end there. A series of statistical studies ensued using different catalogs of QSOs and catalogs of galaxies. The evidence for the physical association continued to accrue with the following:

- Shapley Ames Catalog
- Bright Galaxy Catalog
- Lick Catalog
- IRAS galaxy catalogs
- ROSAT X-ray emissions

A reproduction of the Burbidge list of 40 best cases is shown in Table 3.4.[120] All QSOs lie within 180 arc-seconds of bright galaxies with $m_v < 15.5$. This represents more than 10 times the number of QSOs that would appear at random in the sky. Most prominent are galaxies NGC

QSOs Close to Bright Galaxies ($m_v \leq 15.5$)

Galaxy	m_v	QSO	m_v	z	Separation (″)
UGC 0439	13.8	PKS 0038–019	18.10	1.674	72
NGC 470	12.5	(0117+0317g)	19.8	1.875	93
NGC 470		(0117+0317g)	18.2	1.533	95
NGC 622	14.0	0133+004	18.5	0.91	71
NGC 622		0133+004	20.2	1.46	73
IC 1746	14.5	0151+048	16.91	0.404	6.4
NGC 1073	11.3	BSO 1	19.6	1.945	104
NGC 1073		BSO 2	19.0	0.599	117
NGC 1073		RSO	20.0	1.411	84
NGC 1087	11.5	0243–007	19.1	2.147	170
IC 2402	13.5	0844+319 (4C 31.32)	19.92	1.834	30
NGC 2534	13.7	0809+358	18.5	2.40	121
NGC 2693	13.1	0853+515	19.5	2.31	188
NGC 3067	12.8	0955+326	15.8	0.533	114
NGC 3073	14.1	0958+558	18.8	1.53	144
NGC 3079	11.5	0958+559	18.4	1.154	114
ZW 1022.0–0036	15.5	PKS 1021–006	18.2	2.547	122
NGC 3384	10.8	1046+129	20.6	0.497	149
NGC 3407	14.3	1049+616 (4C 61.20)	16.3	0.422	173
NGC 3561	14.7	1108+289	20.0	2.192	66
NGC 3569	14.5	1109+357	18.4	0.91	31
NGC 3842	12.8	QSO 1	18.5	0.335	73
NGC 3842		QSO 2	18.5	0.946	59
NGC 3842		QSO 3	21.0	2.205	73
NGC 4138	12.1	3CR 268.4	18.1	1.400	174
NGC 4319	13.0	Mk 205	14.5	0.070	43
NGC 4550	12.6	1233+125 (Wdm 8)	17.9	0.728	44
NGC 4651	11.8	3CR 275.1	19.0	0.557	210
NGC 5107	13.8	1319+38	19.5	0.949	40
NGC 5296	14.9	1342+440 (BSO 1)	19.3	0.963	55
NGC 5406	13.1	1358+392	18.5	3.28	95
NGC 5682	14.5	1432+489	19.2	1.940	95
NGC 5832	12.9	3CR 309.1	16.8	0.905	372
NGC 5981	13.9	1537+595	19.0	2.132	10.7
IC 1417	13.6	2158–134	17.8	0.73	76
Anon	15	2237+0305	17.3	1.41	≤0.3
NGC 7465	13.3	2259+157	19.7	1.66	128
NGC 7413	14.8	3CR 455	19	0.543	24
NGC 7714	13.0	PB 5468	18.0	1.871	120

Table 3.4 – 40 close QSO-galaxy pairs

1073 and NGC 3842, each of which has a triplet QSO association with a chance occurrence of 10^{-4}. The observed number of QSO-galaxy pairs increases rapidly as fainter galaxies and fainter QSOs are taken into consideration. This further decreases the chance of random occurrence. In an independent analysis, Zhu and Chu (1995) used the cross-correlation function to provide significant evidence for close QSO-galaxy pairs to the greater distance range of 40 arc-minutes.[121]

The observations of closely aligned QSO-galaxy pairs are a matter of fact. The explanation remains a matter of debate. The only reasonable CH argument comes from Canizares (1981) who detailed the effect of microlensing.[122] Distant quasars that are aligned with individual stars in the galaxy halo experience an intensification of their light, and this makes them more likely to be detected.

This argument fails to explain the abundance of close pairs, however, because the density on the sky of faint QSOs falls short by four or five powers of ten of explaining the abundance of observed pairs.[16] Furthermore, no model of microlensing can explain the brighter close pairs listed in Table 3.4.

The Near Neighbor argument for non-cosmological QSO redshift has one further point requiring clarification. In addition to discordant redshifts, the number of observed QSO/galaxy pairs with the *same* redshifts also continues to grow. These identical redshifts support the older paper (Stockton 1978) that made the argument for CH.* They also support the opposition work (Burbidge *et al.*, 1990), which predicts a QSO intrinsic redshift component that can vary from zero to the largest redshifts measured. Both Stockton and Burbidge listed multiple occurrences of identical redshifts and multiple occurrences where $z_Q > z_G$. Stockton showed the statistical significance of the identical redshifts; Burbidge showed the statistical significance of the discordant redshifts. *Stockton (1978) claims virtual proof of the cosmological redshift hypothesis; Burbidge (1990) claims to disprove the cosmological redshift hypothesis.* An error in reasoning must exist somewhere! It is the older paper (Stockton 1978) that commits the logical fallacy of *false dichotomy*. It assumes without proof that only one of two possible hypotheses can be true. When it correctly disproves LH, it makes the faulty assumption that CH thereby is proved. The Near Neighbor evi-

* See Chapter 2, **Near Neighbor**.

dence, both observational and statistical, now suggests a third possibility for quasar redshift. That leaves the cosmological redshift hypothesis supported by one less pillar.

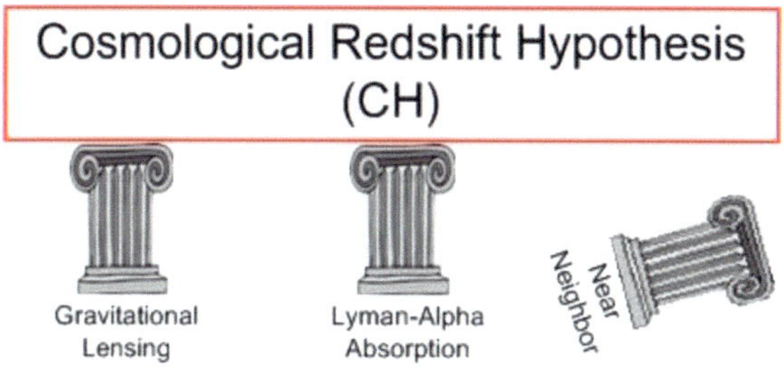

Summary

A good model of the AGN remains elusive. Many observations remain unexplained. There must come a point when one or more of the underlying premises of the model is called into question. The obvious premise to question is that the distances to all AGNs are given by their redshifts, for its rejection would cause many serious problems to vanish. The immediate intention here is not to *verify* that redshift has an intrinsic component. The works of William Tifft and Geoffrey Burbidge serve that purpose. It is instead to show how an AGN model becomes more plausible when the physics behind it can support an explanation of intrinsic redshift that varies from zero to the highest values of *z*. The VH model thereby maintains the existing CH arguments while it removes the difficulties of evolution, iron abundance, X-ray absorption, dust, Faraday rotation, variability, age crisis, and superluminal motion. It also provides a possible explanation for the apparent population of quasars in nearby superclusters and (possibly) quasar redshift periodicity.

A Variable Redshift Hypothesis

By the turn of the century there was general agreement that substantial observational evidence exists showing high-z quasars in proximity to low-z galaxies, many of which are highly active.[123] This creates one of the great mysteries in the field of astrophysics. CH adherents believe it is a visual effect rather than a physical association: dark matter produces a type of gravitational lens that magnifies the faint background of the high-z quasars. This explanation is unsatisfactory. The nature of dark matter remains uncertain, and models do not offer good support for all the required dark lenses. Nevertheless, acceptance of CH remains prevalent, for a great deal of scientific theory rests on the cosmological principle. It will be necessary for any new hypothesis to explain all observations and difficulties shown to date. Quasars with intrinsic redshift alleviate most difficulties discussed in Chapter 3, but they do not explain why some galaxies have the same redshifts as their nearby quasar neighbors. They also do not explain other types of quasar observations, such as the non-existent Hubble-type relationship and the preferred redshift values.

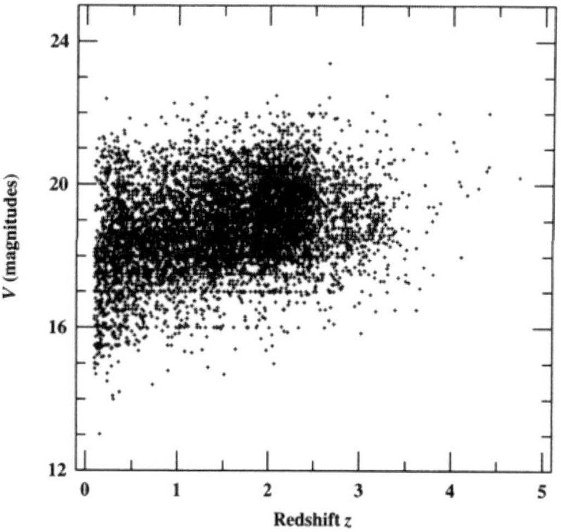

Figure 4.1 – QSO apparent magnitude vs. redshift
© 1997 Cambridge University Press

After the first quasars were observed in the 1960s, it became clear that their apparent spread in luminosity differed from that of normal galaxies. Instead of the smooth relationship observed by Hubble, the plot of apparent quasar magnitude *vs.* redshift is a veritable scatter diagram (see Figure 4.1). Had astronomers seen this scatter before the Hubble plot, it is possible that they would not have concluded so quickly that the universe is expanding.[124] If quasars are cosmological in origin, they must have a very large spread in absolute magnitude during every epoch. If they are closer objects, there must be a variable intrinsic component associated with quasars at *all* magnitudes. In either case, there is something difficult to explain.

Preferred redshifts for quasars have been a subject of debate since the discovery of quasars. Their existence, if proven, suggests new physics at work. In their early book, *Quasi-Stellar Objects,*[94] Geoffrey and Margaret Burbidge disclose that many QSOs have a standard absorbing region with a redshift of 1.95. They suggested that this prevalent redshift value is possibly caused by some intrinsic gravitational shift. Soon afterward studies began to suggest that the redshifts tend to cluster around multiples of some particular value. This became known as *redshift quantization* or *redshift periodicity.** Advocates of CH dismiss such studies as flawed, with no evidence of clustering beyond what is expected due to the distribution of galaxies across the cosmic voids, but they offer no good explanation why the full range of cosmological components z_c does not easily smear out the observed peaks.

The nature of AGNs is shrouded in mystery. Two paths are available to the astronomer in the quest to reveal their secrets. The first path holds steadfast to the cosmological hypothesis (CH) and sets aside any observations that call it into question. The second path diminishes the importance of CH and considers all available observations anew. There is much to suggest that young astronomers should venture along the second path, but that choice risks lack of funding, observing time, and peer support. If travelers avoid the second path, they will miss the opportunity to develop any viable alternative theory that may solve the AGN mysteries. Without a viable new theory, there will remain the firmly held opinion against the existence of non-cosmological redshifts. It's a Catch 22!

Some alternate cosmological models already exist that predict the intrinsic (and quantized) redshift of AGNs. Two notable models are:

---

* See Chapter 3, **Preferred Redshift and Periodicity**.

- Variable Mass – The masses of elementary particles vary in time. The emission of radiation by lower mass particles results in larger redshift. Creation of near-zero mass particles in discrete bursts can cause a quantized distribution of redshift intervals. Variable mass is the linchpin of the quasi-steady state cosmology (QSSC) that predicts pockets of creation occurring throughout time, in contrast to the singular creation event of the Big Bang. The likely form of the 'mini-bangs' is ejection from AGNs whose *young objects* contain particle masses lagging behind the universal mass function. [125] [126]
- Time Dependent Decay – Beginning at the Planck scale, a fundamental unit, U, will decay by a period doubling process, $2^{-D}U$, where D is the index number of doublings that have occurred. Redshift values represent specific decay states from an initial Planck scale value set by the speed of light, c. For example, $D = 13$ yields $2^{-13}c = 36$ km/s, a well-known redshift period. Quantum Temporal Cosmology (QTC) builds upon decay theory to explain observed redshift quantization in terms of the structure of time. [127]

Both QSSC and QTC are well-formulated theoretical models of cosmology built upon the basis of some *new physics* that contradicts the standard interpretation of AGN redshift. Both models make predictions consistent with observations. However, such models are given scant attention based upon the constraint imposed by the cosmological principle. The standard model of cosmology admits various *ad hoc* theories (e.g., Dark Energy), but alternate models are held to the higher standard of building upon *known physics* in order to gain approval. This book shows that it is possible to offer an alternate explanation of AGN redshift by means of known physics.

There is a low statistical probability (10^{-8} to 10^{-9}) that the associations between quasars and low redshift galaxies can be accidental alignments. Also, empirical evidence is consistent with the intrinsic component of redshifts being in discrete amounts and related, such that younger objects have more redshift than older objects at the same distance. [128] On the other hand, there is substantial evidence that some AGNs have no intrinsic redshift whatsoever. It is clear that any new hypothesis must explain such a wide variation in the intrinsic component. It must be a *variable* hypothesis (VH). In order for it to gain support it also must avoid any *ad hoc* new physics. To gain widespread support, it ultimately will have to explain all observations seen to date. Early theories first considered the gravitational field strength of the quasar. Was there anything that could produce enough gravity to explain the observed redshift? Perhaps it is best to re-visit the early gravitational field theories for QSOs.

According to Hoyle

Fred Hoyle (1915-2001)

One of the most creative astrophysicists of the twentieth century was the outspoken Sir Fred Hoyle. Today the scientific world remembers him primarily for his contribution to the theory of stellar nucleosynthesis. He was best known by the public in his early years when, in 1949, he was asked to give a series of talks for the British Broadcasting Corporation. An entire nation heard Hoyle's views concerning the controversy between the theory of the Big Bang and his own cosmological theory of QSSC. His opinion about the nature of the quasars would always be given the highest consideration.

Light leaving the surface of a massive source is redshifted according to the formula

$$1+z_g=\left(1-2GM/rc^2\right)^{-\frac{1}{2}},$$

where G is the gravitational constant, M is the spherical mass, r is the photon's starting distance from M, and c is the speed of light.* Very small spectral shifts, $z_g < 10^{-4}$, are seen in the radiation from white dwarf stars. The first possible explanation for the large redshifts of quasars was a gravitational influence similar to white dwarfs. Two immediate problems of this hypothesis were soon published. Greenstein and Schmidt (1964) showed the difficulty of explaining the sharpness of the spectral lines, [129] and Bondi (1964) computed an absolute upper limit of $z < 0.63$ on the red-

* This assumes the observer is at $R \gg r$, and that the source is non-rotating, spherically symmetric, and uncharged.

shift of light from the surface of any non-rotating equilibrium configuration.[130]

Fred Hoyle held a deep conviction that distant objects in the universe were not appreciably different from those nearby. There must be another explanation for the supermassive objects in deep space. He developed a theory with his frequent collaborator, William A. Fowler, that assumes each QSO to lie at the center of a very massive relativistic neutron star cluster* such that it derives its large redshift from the gravitational field of the cluster. Their theory was published as a letter in a 1967 edition of *Nature*, the prestigious international journal of science. The letter was brief, but the resulting debate about the possible gravitational redshift lasted a long time thereafter. The following paragraph from the *Letter* explains the essence of the model.

> We propose that the model be changed to one in which the observed emission lines come from the centre of an object, not from its surface. Instead of the system comprising a single coherent cloud of gas, we imagine a large number of comparatively compact subunits with radii $\ll R$, the probability for an arbitrary direction through the centre intersecting any of the subunits being small. As an example of the kind of structure we have in mind, we could equate the subunits to clusters of neutron stars or other highly collapsed objects. We could take $M = 10^{12}\text{-}10^{13}\ M_{Sun}$, $R = 10^{18}\text{-}10^{19}$ cm, in which case the probability of an arbitrary direction from the center intersecting a star is $\ll 1$. If the cloud at the centre has radius ~$10^{16}\text{-}10^{17}$ cm, say, light emitted by it passes essentially freely through the star distribution. In this new model the main mass serves to generate a strong gravitational field. Continuum emission is taken to be of non-thermal origin (although faint stars could make a contribution in the near infra-red) while emission lines come from the gas cloud at the centre.[131]

The gravitational redshift theory of Hoyle and Fowler (HF) overcomes the spectral line difficulty as well as the theoretical instability of the massive cluster. However, other research offered a different opinion. In the next few years Edward Fackerell, James Ipser and Kip Thorne (FIT) collaborated on numerous papers that countered HF on the conclusions of relativistic star clusters. For example, Ipser (1969) found that all reasonable models of relativistic spherically symmetric collisionless star clusters become unstable against radial perturbations at values of central redshift

* A relativistic star cluster is a cluster so dense that general-relativistic effects play important roles in structure, evolution, and gravitational redshift.

$z_c > 0.5$.[132] Most interesting in retrospect is the conclusion of Fackerell *et al.* (1969):

> Relativistic gravitational collapse sets in (for $z_c > 0.5$): the stars spiral inward through the gravitational radius of the cluster, toward its center, leaving behind a "black hole" in space with, perhaps, some stars orbiting it. It is tempting to speculate that violent events in the nuclei of galaxies and in quasars might be associated with the onset of such a collapse, or with encounters between an already-collapsed cluster (black-hole) and surrounding stars. Future research may cast light on speculations, just as the recent research, revealing the onset of collapse at $z_c \sim 0.5$ for a wide class of clusters, has cast light (or darkness?) on the Hoyle-Fowler star-cluster model for quasar redshifts.[133]

Research may proceed in different directions after any theory has been refuted. First, other theoreticians may find shortcomings or errors in the refuting opinion. Second, a fundamentally different model may be offered absent the original difficulties. Third, the original model may become amended such that the refutation is overcome. In the case of the HF theory, matters went in every direction.

FIT shortcomings are found in many subsequent publications. Das (1976) found it possible to construct models of static, bound, massive objects in stable equilibrium giving rise to redshifts as high as $z = 4$.[134] Yabushita (1976) showed that the central redshift z_{cen} is an increasing function of the ratio of isothermal gas pressure at the center to the pressure at the core-envelope boundary. QSOs may consist of both gravitational and cosmological effects with a maximum $z_{cen} = 4.43$.[135] Rasio *et al.* (1989) solved general relativity equations for nonlinear dynamical evolution of *collisionless* systems and showed that the cluster remains stable for large central redshift. They concluded that "these clusters provide the first examples of finite, asymptotically flat equilibrium systems which can become arbitrarily relativistic at the center and still remain stable against gravitational collapse."[136] Kogan *et al.* (1998) showed that dense star clusters are stable against relativistic collapse up to infinite values of the central redshift.[137]

All objections to the FIT analysis notwithstanding, the model of the relativistic star cluster was becoming a dying ember. The possibility of any intrinsic gravitational redshift had waned once astrophysicists concluded that the requisite large masses did not exist. Furthermore, the golden age of black hole theory had dawned to inflame the passions of the astrophysical community.* Salpeter and Zel'dovich (1964) conceived the idea that quasars were gigantic black holes with infalling gas streams that collide and produce enormous radiation. Lynden-Bell (1969) further argued that centrifugal forces make the streams spiral around the hole, forming an accretion disk. It wasn't very long until astrophysicists came to accept the idea that the brilliant light from the center of 3C 273 was probably a friction-heated accretion disk.†

This model still would need to consider the effect of the large gravitational field of the supermassive black hole (SMBH). Light is affected in a gravitational field. It will appear redder since its characteristic frequency decreases. It will appear dimmer since the number of photons emitted per second is reduced. What, then, formed the basis to downplay the influence of SMBH gravity on emitted radiation? Recall that the astronomical community considered the redshift debate closed by 1983.‡ Some references did exist on the influence of the gravitational field near the black hole event horizon. Consider the words of Silk (2005):

> The gravitational redshift due to a supermassive black hole may have already been detected. Active galactic nuclei are thought to be powered by supermassive black holes. Emission in X-rays from iron is observed to be redshifted from the expected frequency by about 10 per cent. [139]

Overall, however, those who theorize AGN structure and evolution treat such effects as minimal and consider CH unassailable.

Fred Hoyle held steadfast to his belief that gravitational forces are the key to understanding the nature of QSOs. Fellow researchers had come to rescue his theoretical model of the dense neutron star cluster, but observations suggested that such a phenomenon does not exist in nature. He opted to amend his model with a completely different type of massive particle, the magnetic monopole, with predicted mass of 10^{18} larger than a proton. An isolated magnet with only one magnetic pole has never been observed in nature, but arguments for its existence come from grand unification theories (GUTs) and developments in the field of quantum gravity.

---

* See Kip Thorne's *Black Holes and Time Warps* for an excellent history of black hole theory and its theorists. [138]

† See Chapter 1, **The Third Question: Do Some Extragalactic Objects Exhibit Intrinsic Redshift?**

‡ See introduction to Chapter 2.

Hoyle theorized that QSOs are powered by magnetic monopole annihilation in...

> ...a nested arrangement of potential wells, with the emission spectrum coming from gas accumulated at the bottom of the central well. Material falling radially into such a system accelerates down the curved wall of the outermost well until it reaches the nearly flat bottom, where it coasts for a while until it reaches second-outermost well. The material then accelerates to a greater speed down the wall of the second well, until it reaches the bottom of that well, where it again coasts for a while until it reaches the third-outermost well. And so on to the bottom of the central well. The redshift z_{em} of light received by a distant observer from emitting gas at the bottom of the central well is given by
>
> $$1 + z_{em} = (1 + z_1)(1 + z_2)...(1 + z_n)$$
>
> where z_1, z_2,... z_n are the separate redshifts of the potential wells which form the nest.[140]

All previous difficulties dissolve under the magnetic monopole model. The collisional issues of dense star clusters disappear. Variations of emission line widths are an effect of the variable gravitational wells. Monopole annihilation produces the strong high-energy (UV, X-ray, γ-ray) emissions. Very rich absorption lines are more likely found when there is a large emission-line redshift. Consider his explanation for the Lyman-α forest lines:

> Instead of thinking of material falling from outside to the centre, it is interesting to ask at what places within such a structure might material remain in equilibrium. Obviously at the bottom of the central well, but also very nearly on shelves near the bottom of each successive well – the shelves where infalling material coasts for a while. Numbering the wells from the outside, material accumulated on the shelves could produce absorption lines in radiation emitted by gas at the bottom of the central well, the absorption redshift z_{abs} as seen by an external observer having a possible series of values,
>
> $$1 + z_{abs} = 1 + z_1, (1 + z_1)(1 + z_2),...,(\sim 1 + z_{em}).$$

In the end, any scientific theory must adhere to the observations. The magnetic monopole model proffered by Hoyle stands strong in this regard with one glaring exception, no monopole has ever been observed. The SMBH model suffers from many unexplained anomalies as outlined in Chapter 3, but it rests easier on the basis that black holes are an (indirectly) observable entity. Observational evidence for the existence of black holes has grown steadily through the years, and observatories are beginning to explore their gravitational wave emissions. Abbott *et al.* (2016) announced the discovery of a gravitational wave signature of a binary black hole merger.[141] Two initial black holes with masses of 36 M_{Sun} and 20 M_{Sun} (at $z = 0.09$) were observed to coalesce into a final black hole with mass of 62 M_{Sun} with 3 $M_{Sun}c^2$ radiated in gravitational waves.*

There is nothing conclusive about either the SMBH model (with associated accretion disk) or Hoyle's magnetic monopole model. Perhaps some combination of these models may prove better. Was Fred Hoyle correct in his assumption that gravitational forces best describe the enormous power of the QSO? At a minimum it is necessary to understand all gravitational factors before that question can be answered with any confidence.

Proximity and Mass of the Black Hole

Most AGN spectral line profiles show shift, broadening and asymmetry. The widely held explanation is a Doppler effect caused by clouds of ionized gas moving independently with high velocity dispersion. Any effect by the gravitational field of the black hole has been considered negligible and mostly ignored. There does come a point, however, where the effect can become appreciable. The wavelength of light, λ_0, in a gravitational field shifts by

$$\Delta\lambda = \lambda_0 \Delta U / c^2 ,$$

where c is the speed of light and ΔU is the difference of the gravitational potential between the points of emission and observer. Photons cannot emit from a black hole, but they do emit from the surrounding gaseous clouds. For light emitted from a nearby cloud, the wavelength shifts by

$$\Delta\lambda = \lambda_0 GM / rc^2 ,$$

where G is the gravitational constant, M is the mass of the black hole, and r is the distance from the black hole to the cloud. This assumes that the gravitational potential is negligible at the point of observation. The equation states that redshift will depend upon the proximity and mass of the black hole. It also states that redshift increases as a function of the initial wavelength. Popovic *et al.* (1995) provided a theoretical calculation of the non-negligible influence on spectral lines when the emission region is 100

---

* See later section, **Gravitational Waves**, for an explanation of the GW signature of a binary merger ringdown.

R_S from a SMBH of $10^7 - 10^8$ M_{Sun} (where R_S is the Schwarzchild radius). They concluded: "the gravitational effect may become observable and should be taken into account in analyses of spectral lines originating in sufficiently dense regions around massive nuclei of active galaxies." [142]

Michael R. Corbin (1995) put this to the test by investigating the known asymmetries in broad $H\beta$ and $C_{IV}\lambda1549$ emission lines. Using the $z < 0.5$ data compiled by Boroson and Green (1992), [143] he first found a strong correlation between $H\beta$ asymmetry and soft X-ray (2 keV) luminosity. He then proceeded to investigate the origin of the $H\beta$ and C_{IV} asymmetries by numerical analysis of different broad-line region (BLR) models. Evidence suggests that the traditional BLR consists of an intermediate-line region (ILR) composed of optically thick clouds and a very broad line region (VBLR) of optically thin material lying closer to the central continuum source. A likely model for the BLR would be that the asymmetries arise from a relative shift of the velocity centroids of the ILR core and the VBLR $H\beta$ (or C_{IV}) base component. Indeed, Corbin (1995) [144] found this model to be the best fit for the observed data. Histograms 4.1 and 4.2 show the $H\beta$ redshift distributions of the core and base components.

The parameters *z*(core) and *z*(base) are the redshifts of the ILR and VBLR centroids measured relative to the system velocity defined by the position of $O_{III}\lambda5007$. These results indicate that the asymmetries can be modeled with a two-component BLR where the VBLR centroid shifts relative to systemic velocity. Relying on the fact that the $H\beta$ redshift has a strong correlation to soft X-rays (which in turn trace the mass of the SMBH), Corbin posits that the $H\beta$ photons must experience a gravitational redshift in some proportion to the mass and distance from the black hole. Although his research is consistent with the predictions of Popovic *et al.*

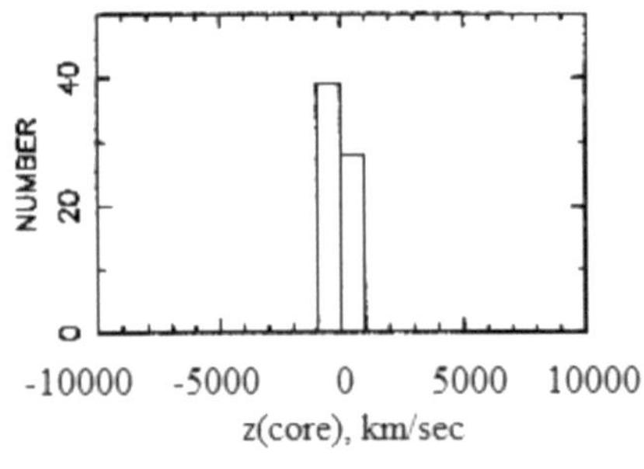

Histogram 4.1 – ILR redshift distribution

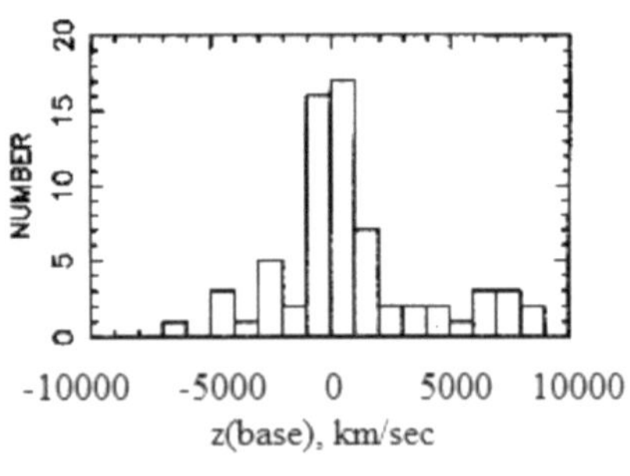

Histogram 4.2 – VBLR redshift distribution

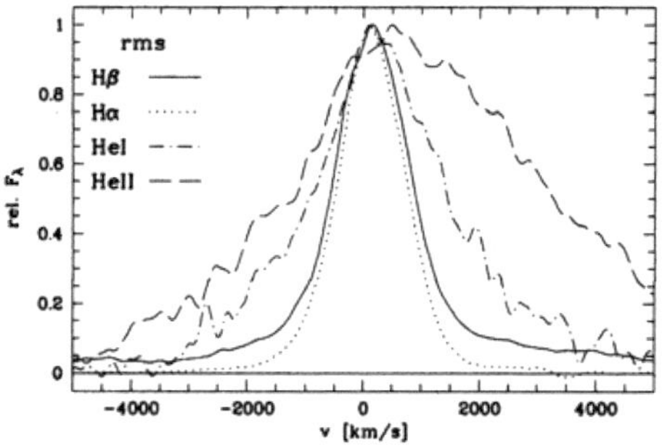

Figure 4.2 – Line profiles of Hα, Hβ, HeI and HeII
© 2003 EDP Sciences

Line (1)	FWHM(rms) [km s⁻¹] (2)	τ_{cent} [days] (3)
HeIIλ4686	4444±200	3.5^{+2}_{-2}
HeIλ5876	2404±100	10.8^{+4}_{-4}
Hβ	1515±100	23.5^{+4}_{-4}
Hα	1315±100	32.5^{+4}_{-4}

Table 4.1 – FWHM vs. distance to black hole

(1995), it does not present conclusive evidence. VBLR kinematics remains uncertain, and any significantly thick or flattened VBLR will have some effect on the observed line profiles. Further substantiation came when it was later discovered that even stronger redshift occurs from sources that originate closer to the black hole. In an analysis of the variations in the Seyfert 1 Mrk 110 ($z = 0.03552$) spectra, Kollatschny (2003) compared the emission lines of $H\alpha$, $H\beta$, $He_I\lambda5876$, and $He_{II}\lambda4686$.[145]

Figure 4.2 shows the root-mean-square (rms) line profiles in velocity space. Table 4.1 provides the FWHM and the computed light travel time from the emitting source to the central black hole. It is readily seen that successively broader (and redder) line profiles are closer to the black hole. Other studies show the $H\beta$ redward asymmetries of dozens of quasars are well correlated with the black hole mass.[146][147] The outstanding question is whether there is sufficient *cold* gas close to the black hole that is able to emit the very luminous redshifted emissions. If the temperature is too high, the gas remains fully ionized. Bon *et al.* (2015) estimated gas temperatures at 100 R_S from a 10^9 M_{Sun} black hole for two different geometries of the accretion disk.[148] They found the gas temperature to be much too high, $T_e \sim 4 \times 10^6$ K, for an optically thin advection dominated accretion flow (ADAF), but it is three orders of magnitude less for a geometrically thin, optically thick disk.* Gravitational redshift is possible for *some* geometric configurations of accretion disks.

Everything comes into sharper focus with the theoretical analysis of Muller and Wold (2006).[149] In their study of a large range of distances to

* See Chapter 5, **Accretion Disk Geometry**, for discussion of different accretion disk geometries.

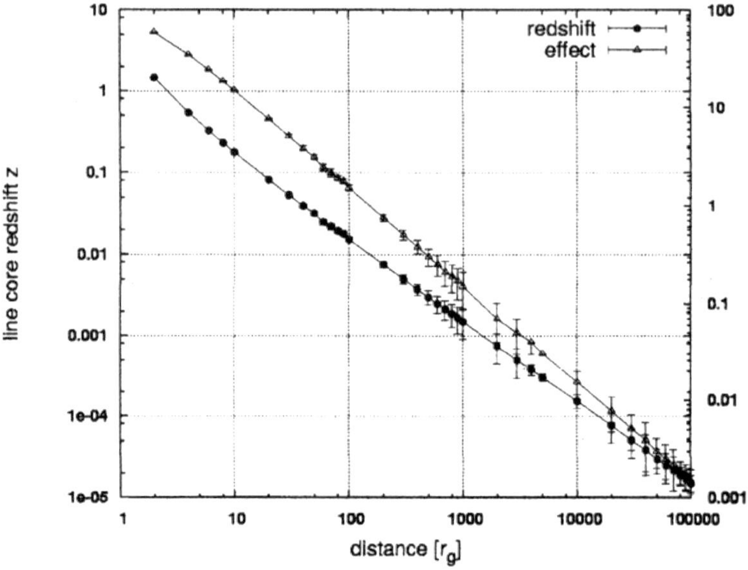

Figure 4.3 – Redshift vs. radial distance to black hole (accretion disk inclined to 1°)

the black hole and observer viewing angles, they were able to quantify the influence of Doppler *vs.* gravitational redshift on BLR emissions. The data show quite convincingly that the lines are subjected to ever stronger gravitationally induced shift and distortion as the emitting region moves closer to the SMBH.

Figure 4.3 displays two plots as a function of the gravitational radius, $r_g = GM/c^2$ ($= R_S/2$ for a Schwarzchild BH), to a black hole with an accretion disk inclined 1° to the observer. The core redshift of emission lines is given by the left *y*-axis, plotted by the lower line of filled circles. The percentage of the gravitational redshift is given by the right *y*-axis, plotted by the upper line of triangles. A gravitational strength of 100% occurs at the event horizon of the black hole.

The gravitational effect on redshift becomes noticeable (> 0.2%) when the accretion ring is inside a radius of $1000r_g$, and the effect becomes sizable inside $2.5r_g$. In particular, notice that an AGN at an actual cosmological distance of $z_c = 1$ with an accretion disk inside a radius of $2.5r_g$ will have a total observed redshift of $z_{tot} > 2$. This fundamentally disrupts the cosmological redshift computation.

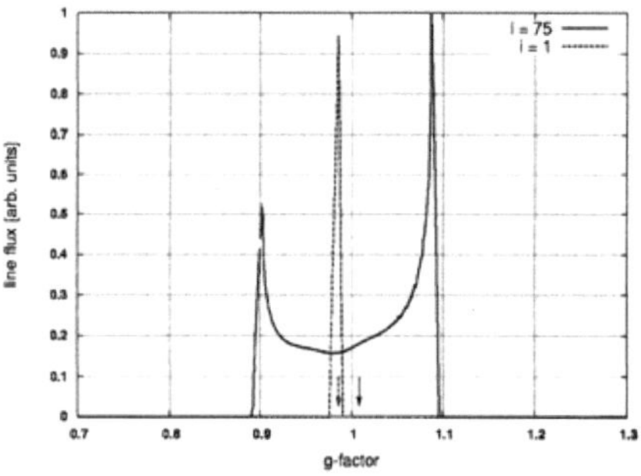

Figure 4.4 – Line broadening
(disk inclined to 75°)
© 2006 EDP Sciences

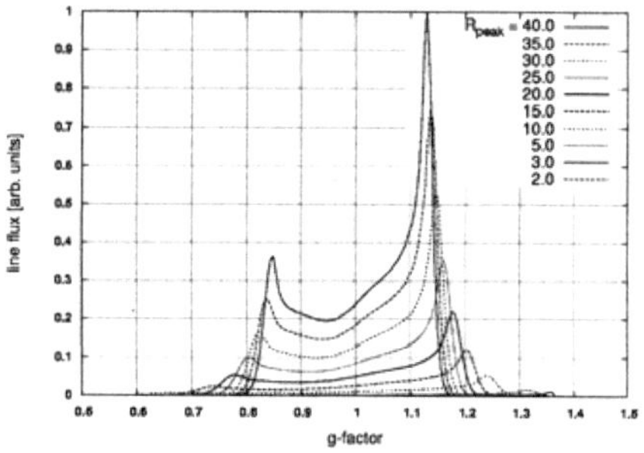

Figure 4.5 – Line flattening
(disk approaches BH)
© 2006 EDP Sciences

Another variable of observed redshift is the angle at which the accretion disk is inclined to the observer. The orbital motion of disk material causes a broadening of the emission flux. Figure 4.4 shows the noticeable line broadening that occurs if a disk at $r_g = 100$ is rotated from face-on to the highly inclined angle of $i = 75°$, where the indicated g-factor = $1/(1+z)$.

The flux also becomes more and more suppressed as the distance of the accretion disk moves closer to the black hole. Figure 4.5 shows the effect on successively smaller values of r_g, while maintaining the inclination angle of $i = 75°$. For an accretion disk at radius of $1r_g$, the emission line disappears altogether.

The equation $\Delta\lambda = \lambda_0 GM / rc^2$ shows that wavelength (and hence redshift z) varies as a function of the distance to the black hole and its mass. Gravitational redshift from an inclined accretion disk also causes a suppression of flux in a manner that the emission line becomes asymmetric and skewed. We should thus write at least three independent factors in the function for the gravitational redshift, z_{grav}.

Observational evidence suggests that there are errors in some AGN distance measurements. This can happen when a quasar has an accretion disk 'close' to the black hole, but evidence of such proximity is not easily attained given the challenge of probing gravitationally redshifted spectral lines. The analyses of Popovic *et al.* (1995), Corbin (1995), Kollatschny (2003), Muller and Wold (2006), *etc.* are tantalizing. They help provide an explanation for some inaccuracies and can thereby serve as a building block for a plausible VH of AGN redshift. However, the theoretical instability of rings within $2.5r_g$ strongly suggests that additional factors are necessary for a complete description of any significant gravitational redshift.

$$1 + z_{grav} = f\,(M_{BH}, r, i, x_1, x_2, \ldots, x_n)$$

M_{BH} is the mass of the black hole

r is the radial distance of the accretion disk from the black hole

i is the angle of inclination of the accretion disk to the observer

$x_1, x_2, \ldots, x_n$ are other factors under consideration

Black Hole Rotation (Spin)

In addition to the mass, M_{BH}, any proper characterization of a black hole requires evaluation of J, the angular momentum.* Accretion rate, radiation efficiency, and extractable energy all depend on the angular momentum. The rotation of the black hole produces some amount of gravitational redshift, which is difficult to determine precisely. The study of gravitational effects on rotating neutron star emissions provides some insight into the amount of redshift that may occur from a quasar. Pulsars are neutron stars with rapid rotation rates in the thousands of radians per second. The emissions appear 'pulsed' due to the fact that the radiation is seen only when directed toward Earth. The large mass and angular velocity of the pulsar make it an ideal candidate for the study of general relativistic effects.

Dubey and Sen (2015) provided mathematical proof that the gravitational redshift of any photon emitted from the surface of a rotating sphere depends on the spherical coordinates of the emission location as well as the mass and angular velocity of the sphere. [150] Figure 4.6 shows the calculated values for pulsar PSR B 1913+16 with known angular velocity, $\Omega = 106.4$ rad/s. The gravitational redshift of the Lyα line is plotted against the azimuthal angle, ϕ, at fixed polar angle, $\theta = 90°$. The light (and z_g) of the pulsar is typically observable only when the beam is directed toward Earth at $\phi = 0°$. The redshift also varies as a function of the polar angle (as shown by similar argument of Muller).† Notice the significant redshift of the photons as they pass through the equatorial maxima, $\phi = 90°$ and

* Complete characterization requires mass, angular momentum, and electric charge Q, but electric charge is assumed negligible in most studies.

† See previous section, **Proximity and Mass of the Black Hole**.

$\phi = 270°$. The Sun also produces some high redshifted photons as it instantaneously passes through these points. The effect of gravity on PSR B 1913+16 is confirmed by observation.[151] Whether similar effects exist for a quasar is the germane issue. Light from the accretion disk must be sufficiently close to a black hole with large angular momentum in order to produce gravitational effects. While the mass of a black hole can be measured by interactions with gas and other materials,* evaluation of the requisite angular momentum and distance to accretion disk are both very difficult to determine.

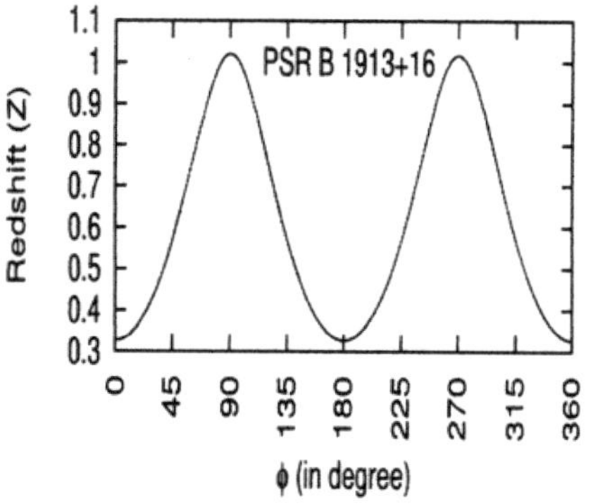

Figure 4.6 – z_g vs. azimuth angle

The terms *spin* and *spin parameter* are used frequently when referencing black hole angular momentum. Unfortunately, usage is quite inconsistent in the technical journals. Depending on the author(s): *spin parameter* is defined as the ratio *J/M*, the percentage of maximum angular momentum, or some other dimensionless value α, where $-1 < \alpha < 1$; *spin* refers to the angular momentum or the defined spin parameter. In order to provide consistent meaning in this book, spin, *S*, and spin parameter, α, are defined as:

$$S = \alpha J_{\max},$$

where $-1 < \alpha < 1$ is the (dimensionless) spin parameter, and

$$J_{\max} = M r_{ms} c = M\left(GM/c^2\right)\cdot c = GM^2/c.$$

The spin of a black hole produces the bizarre effect of 'dragging' space itself into circulation. This produces added centrifugal force and allows gas (in prograde motion) closer to the black hole event horizon where temperatures (and luminosities) reach maximum values. For a black hole spinning at the maximum possible rate, gas moves near $1r_g$ before making the plunge into the hole at the speed of light. As the location of emitted

* See Chapter 5, **Black Hole Mass Estimations**, for discussion of the preferred methodology for determining black hole mass.

α	r_{ms}/r_g
−1.0	9.0
0	6.0
0.1	5.67
0.5	4.23
0.9	2.32
0.998	1.24
1.0	1.00

Table 4.2 – Spin parameter vs. r_{ms}

photons approaches $1r_g$, the gravitational redshift goes to infinity. (See Figure 4.3).

The marginal stability radius, r_{ms}, also known as the innermost stable circular orbit (ISCO), is the location beyond which particles lose their orbital motion and fall into the event horizon. General relativity enables the calculation of r_{ms} as a function of α.

Some important values of r_{ms} and associated α are shown in Table 4.2. The last entry shows that $r_{ms} = r_g$ for a maximally spinning black hole ($\alpha = 1$). A non-rotating hole ($\alpha = 0$) has an ISCO of $r_{ms} = 6r_g$, and an accretion disk in retrograde motion to a maximally spinning black hole ($\alpha = -1$) has $r_{ms} = 9r_g$. The table also shows that black holes spinning at $\alpha = 0.9$ support a stable orbit at $r_{ms} = 2.32r_g$. This value is inside the zone of significant gravitational redshift per Muller and Wold (2006). It hence becomes crucial to understand if, and how many, black holes spin up to the critical value of $\alpha = 0.9$ that can maintain a highly redshifted accretion disk.

Theory predicts that black holes will evolve by two principal means: binary coalescence and/or accretion from matter orbiting the hole. A binary coalescence unites two separate black holes into one body or mass. When a black hole unites with a binary companion of comparable mass, the net spin will increase. When the companion is smaller, the net spin will decrease. Accretion, however, always tends to spin holes up efficiently, independent of the orientation of black hole and accretion disk. A non-rotating hole becomes maximally rotating after a gain of 240% of its initial mass. Theorists have used knowledge of black hole evolution to develop models that predict the relative distribution of spin throughout (ΛCDM) cosmic history. Several such predictions from the early model of Volonteri *et al.* (2005) are revealed in Figures 4.7 and 4.8.[152] Solid histograms represent cases where the seed black hole is assumed to have an initial spin parameter of $\alpha = 0.6$. The dashed histograms represent non-spinning seed holes with the initial spin parameter of $\alpha = 0$.

The left panel of Figure 4.7 depicts the effect of coalescences without any accretion events. The spin distribution maintains its initial centroid value. The spread in distribution over time occurs from the occasional

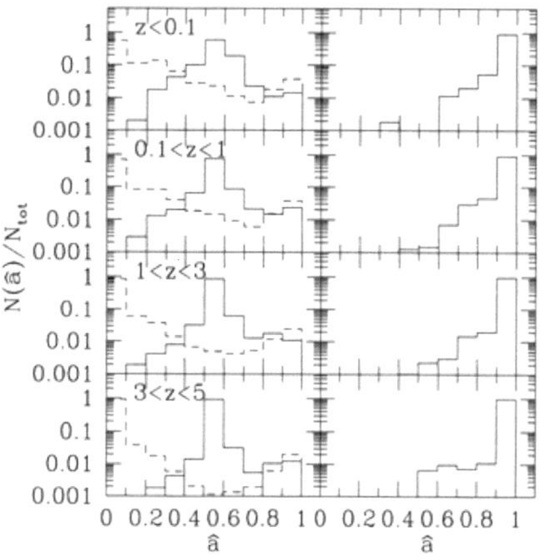

Figure 4.7 – Spin distribution without and with accretion

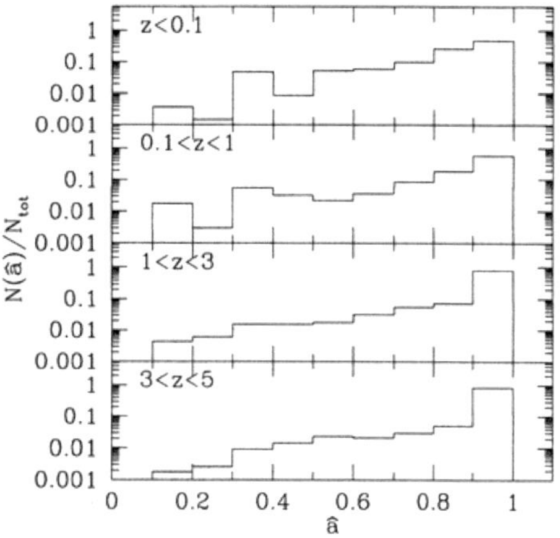

Figure 4.8 – Spin distribution *via* thick disk

small companion that 'finds' the larger hole. The right panel depicts the combined effect of coalescences combined with accretion events. The majority of holes soon attain a value of $\alpha > 0.9$ that is not relinquished over time. Both of these scenarios assume a thin accretion disk.* It is possible that much of the high-*z* black hole growth occurs *via* a radiation pressure supported thick disk that produces slower spin-up due to the longer timescale for the black hole axis to align with the angular momentum. Figure 4.8 depicts the effect of geometrically thick accretion disks on the spin parameter. Notice the high density of spin parameters at critical values $\alpha > 0.9$ at every epoch.

The worthiness of any theoretical model is only as good as the accuracy and breadth of the assumptions on which it is built. It takes only one missing or incorrect supposition to invalidate the entire model. The Volonteri model assumes some degree of angular momentum 'memory' by means of *coherent accretion*. Matter falls inward without significant interruption during one continuous accretion event. If the model were built instead on the assumption of *chaotic accretion*, the model would point to dramatically different values for the spin parameter. Chaotic accretion consists of a long series of low-mass accretion episodes, each of whose angular velocity vector is chaotically oriented from the next. The co- and counter-alignment of these vectors to the black hole spin thereby thwarts the possibility of maximal spin-up. King *et al.* (2008) modeled discontinu-

* See Chapter 5, **Accretion Disk Geometry**, for discussion of thin and thick disks.

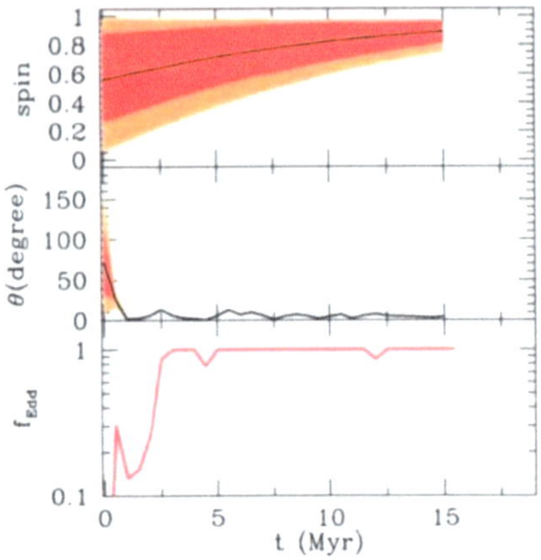

Figure 4.9 – Dynamics of accretion event

ous, randomly oriented accretion with each episode on scales ~10^3 times smaller than the total mass of the entire event. The outcome is a black hole whose spin parameter rapidly adjusts to an average value $\alpha \sim 0.1$-0.3.[153] Note that their model assumes thin accretion disks. When the inefficient cooling of geometrically thick disks is taken into account, the conclusions no longer hold.

The behavior of infalling matter and angular momentum became a key issue for subsequent models. The effects of star formation, stellar feedback, radiative transfer, and metal pollution were introduced to gain further insight into the spin evolution and other associated accretion dynamics. Simulations showed that the activity occurs in three distinct phases. The first phase consists of 'healthy' infall of material into the black hole from a circumnuclear disk. The second phase consists of a 'blowout' of the material resulting in a final 'starvation' phase. Analysis shows that local star motion and feedback effects can create a partial break in the general coherency of the infall, but the energy is insufficient during the healthy phase to randomize the orbits into a chaotic event.* The net result is a gain of high spin for the black hole during the healthy and blowout phases. Maio *et al.* (2013) calculated what happens when all these parameters are put into three-dimensional, $\mathcal{N}$-body, hydrodynamical, chemical simulations. Figure 4.9 is representative of the evolution of spin magnitude (top panel), relative angle between black hole spin and disk orbital angular momentum (middle panel), and corresponding luminosity (bottom panel).[154] Different initial alignment, star formation rates, radiative cooling, metal pollution rates, and initial spin do not make any significant difference. Angle alignment occurs within several Myr; accretion rate reaches a maximum within 5 Myr; and the spin parameter exceeds the critical value of 0.9 within 15 Myr. Volonteri *et al.* (2013) improved their model

* However, chaotic accretion can occur in cases of disk instabilities or massive clumps of cold gas.

with MBH-MBH mergers, merger-driven gas accretion, stochastic fueling of MBHs through molecular cloud capture, and a basic implementation of accretion of recycled gas.[155] Figure 4.10 reveals the dramatic impact on α by increased black hole mass. Green triangles indicate $10^6\ M_{Sun} < M_{BH} < 10^8\ M_{Sun}$. Orange squares indicate $M_{BH} > 10^8\ M_{Sun}$. Left boxes show chaotic accretion; right boxes, cohesive accretion; upper boxes, accretion at normal rates ($f_{Edd} > 0.1$); lower boxes, accretion at very low rates ($f_{Edd} > 0.001$). The critical spin parameter value of 0.9 is attained only for $M_{BH} > 10^8\ M_{Sun}$ and only at $z > 0.5$. Lower mass black holes are mostly insensitive to the chaotic and cohesive models, and all differences disappear at $z < 0.5$.

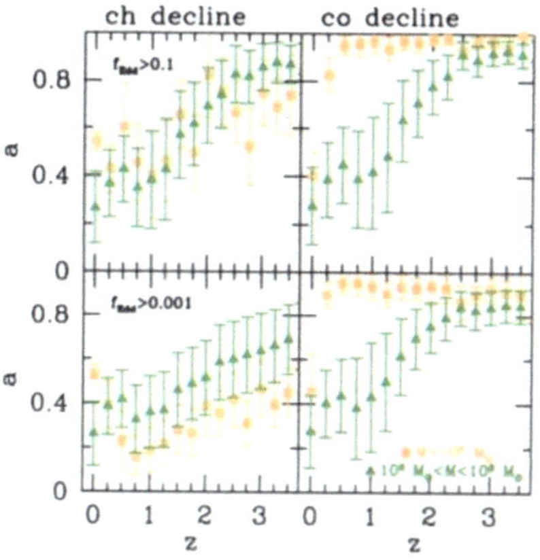

Figure 4.10 – Spin parameter vs. M_{BH}

Another theoretical approach to determining the spin parameter uses the calculated efficiency, η, of converting rest mass to electromagnetic radiation due to infall of mass m from infinity to r_{ms}. In general, $\eta = \left[E(\infty) - E(r)\right]/mc^2$. The subsequent increase in M_{BH} becomes $(1 - \eta)m$. Table 4.3 shows some values of α as a function of η (and r_{ms}).

Given that light derives from accretion of matter onto a black hole, the radiative efficiency can be calculated from the formula $\eta \approx \Delta\varepsilon/\Delta\rho c^2$, where $\Delta\varepsilon$ is the increase of the radiative energy and $\Delta\rho$ is the increase of black hole mass density. Using *i*-band magnitudes and broad Hβ and Mg_{II} lines from the SDSS database to calculate the luminosity and mass functions, Wang *et al.* (2006) determined the efficiencies (η) for the redshift range of $0.4 < z < 2.1$. The average radiative efficiency was found higher than 30%, or $\alpha > 0.9$.[156]

α	r_{ms}/r_g	η
–1.0	9.0	0.038
0	6.0	0.057
0.1	5.67	0.061
0.5	4.23	0.082
0.9	2.32	0.156
0.998	1.24	0.321
1.0	1.00	0.423

Table 4.3 – spin parameter vs. η (and r_{ms})

Predictions of high spin are not limited to theoretical models of black hole evolution. Further support that most massive black holes are spinning rapidly comes from Cao (2007) who finds that both the observed intensity and spectral shape of the X-ray background (XRB) are best reproduced with the adoption of black holes having a spin parameter of $\alpha = 0.9$. [157]

Progress in the research of active galaxies depends heavily on the means to measure the angular momentum of an individual black hole. Neither the change in accretion disk luminosity nor an increase in M_{BH} is measurable over available time frames. It becomes necessary to measure the marginal stability radius, r_{ms}, as the only available means to determine the spin. This became possible with the high resolution X-ray data from Chandra, XMM-Newton, Suzaku and NuSTAR. These observatories are capable of detecting high-energy radiation reflected off material at the inner edge of the optically thick accretion disk. Most prominent in the observed broadband X-ray spectra is the Kα iron line (6.4 keV) with a blue-shifted peak and a highly redshifted wing. The widened effects of the iron line are associated with the relativistic frame dragging of the ISCO around a rapidly spinning black hole. Tanaka *et al.* (1995) found the first robustly detected AGN broad iron line from the Seyfert galaxy MCG 06-30-15 at redshift $z = 0.015$. It has a clear profile with an asymmetric Doppler shift of ~100,000 km s^{-1} and line flux that drops sharply above the rest energy of 6.4 keV. [158] This is consistent with a face-on accretion disk near the central black hole. Subsequent observations of emission line variability and its time-lag relative to variations in the continuum emission intensity made it possible to constrain the spin parameter to $\alpha = 0.989$ at 90% confidence. [159] The higher values of α translate to smaller values of r_{ms} (see Table 4.3), which produce a higher incidence of significant gravitational redshifts predicted by Muller and Wold (2016). Everything thereby suggests that the spin parameter is another important factor for the gravitational redshift.

$$1 + z_{grav} = f\,(M_{BH}, r, i, \alpha, x_1, x_2, \ldots, x_n)$$

M_{BH} is the mass of the black hole

r is the radial distance of the accretion disk from the black hole

i is the angle of inclination of the accretion disk to the observer

α is the spin parameter

$x_1, x_2, \ldots, x_n$ are other factors under consideration

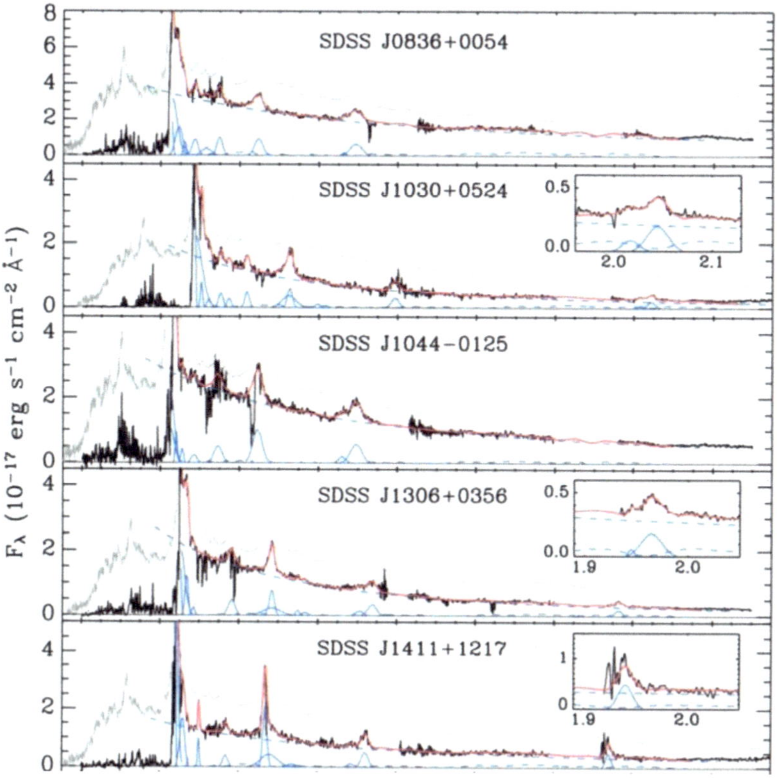

Figure 4.11 – Near-infrared spectra for five quasars (z = 5.82 to 6.28)

Although the early papers [142] [144] [145] [146] [147] [149] on non-negligible gravitational redshift were augmented by later modeling and indirect observations of black hole spin, many observed phenomena remain for which there are no adequate explanations. Most difficult to explain are the spectra of high-*z* quasars. A fundamental property of quasar evolution is a lack of change in spectral properties throughout cosmic time. The spectral composite of high-*z* quasars is almost identical to the low-*z* composite in terms of emission line strength, spectral slope, flux ratios (normalized to the C_{IV} fluxes), and signs of metallicity. This is not easy for CH to explain. But the dense LAFs at high redshift present a difficulty for any VH explanation on the sole basis of the gravitational attraction of supermassive, rotating black holes. Consider the near-infrared spectroscopy by Jiang *et al.* (2007), which compares the spectra of five high-*z* quasars to a low-*z* composite spectrum. Figure 4.11 shows the spectra of the quasars in the redshift range of $5.82 < z < 6.28$. [160]

The red lines indicate the sums of all the components, and the blue lines indicate best fits. The low-redshift composite gray line is overlaid for the sake of comparison. The Lyα emission spikes are seen near the left end of the diagrams. Note how the spectral features redward of Lyα (right of the spike line) have not evolved from the low-z composite for any of the quasars. Note also the heavy absorption blueward of Lyα that is not seen in the low-z composite. The strong LAFs in the spectra are a signature of high-z quasars. A robust VH will require additional factors to explain this phenomenon.

$\mathcal{N}$-Body Gravitation

A black hole is only one of the tens of trillions of objects within a galactic system. A thorough analysis of the gravitational potential inside the galaxy requires an understanding of the net effect of all these objects. For a normal galaxy the effect on electromagnetic radiation will be negligible. Consider the Milky Way which has an estimated 400 billion stars and approximate diameter of 1-1.8×10^5 light-years. The maximum average wavelength shift of light leaving the bottom of the gravitational well is ~10^{-2} Å. It is easy to understand why astronomers consider cosmological expansion as the cause for the large redshifts of the galaxies. If stars produce any noticeable Newtonian gravitational reddening of light, they must be far denser and/or relativistic than what is observed in a normal galaxy.

Stars are neither static nor uniformly dense within a galaxy. Their relative motions produce a non-stationary gravitational field that in turn alters the frequency of any observed radiation. Until 1990 the working assumption was that the biggest influence on light rays occurs as the photons pass near a massive static body. This assumption fails if the light-deflecting body moves too fast in a specific direction causing a weak gravitational field to become strong in a chosen coordinate frame.[161] Kopeikin and Schafer (1999) calculated the (general relativity) metric tensor to first post-Minkowskian order (1PM) to offer a solution to the gravitational shift of electromagnetic frequency in an arbitrarily (translational) moving $\mathcal{N}$-body system.[162] Constant point-like masses, m, position vectors, $\mathbf{x}_n$, and velocity vectors, $\mathbf{v}_n$, ($n = 1,2,\ldots,\mathcal{N}$), are the assumed known quantities in a recursive equation that solves $1+z=\nu_0/\nu$, where ν_0 is the emitted frequency and ν is the observed frequency. For various simulations, Klioner and Peip (2003) showed that the assumption of stationary bodies leads to errors of 1 mas in light deflection, whereas the post-Minkowskian model predicts appreciably smaller errors of ~0.002 μas deflection.[163] In addition to its mass and translational velocity, the light-deflecting body will modify electromagnetic radiation by virtue of its rotational motion. Kopeikin and Mashhoon (2002) extended the post-Minkowskian light-propagation formalism to include the spin-dependent terms of the light-deflecting bodies. Spin is integrated into the energy-momentum tensor $T^{\alpha\beta}(t,x) = T^{\alpha\beta}_{\text{Mass}}(t,x) + T^{\alpha\beta}_{\text{Spin}}(t,x)$, where $T^{\alpha\beta}_{\text{Mass}}$ and $T^{\alpha\beta}_{\text{Spin}}$ are tensor components generated by, respectively, the mass and spin of the body, and t and x are the time and spatial coordinates of the underlying inertial coordinate system.[164]

The computational burden of including all these interactions in a large $\mathcal{N}$-body system becomes prohibitive. The typical approach is to include the effects of the black hole on each star (*i.e.*, $\mathcal{N}$ two-body systems) and consider other interactions negligible. There now is reason to believe that this approach is insufficient to estimate the full gravitational potential. Will (2014) showed the necessity of including post-Minkowskian 'cross terms' in the equations of motion.[165] While the effect of other stars may appear small at first, the effect becomes 'boosted' over relativistic time and leads to unforeseen consequences in the longer term.* Thorough assessment of gravitational redshift requires that all $\mathcal{N}$-body factors are taken into consideration.

$$1 + z_{grav} = f\,(M_{BH}, r, i, \alpha, m, \mathbf{x}_n, \mathbf{v}_n, \mathbf{T}_n)$$

M_{BH} is the mass of the black hole

r is the radial distance of the accretion disk from the black hole

i is the angle of inclination of the accretion disk to the observer

α is the spin parameter of the black hole

m is the pointlike mass of each member in the $\mathcal{N}$-body galaxy

$\mathbf{x}_n$ is the positional vector of the nth member in the $\mathcal{N}$-body galaxy

$\mathbf{v}_n$ is the velocity vector of the nth member in the $\mathcal{N}$-body galaxy

$\mathbf{T}_n$ is the spin tensor of the nth member in the $\mathcal{N}$-body galaxy

---

* Fred Hoyle might have relaxed his constraints on the (HF) nuclear star cluster model if he had the benefit of these *N*-body simulations. See previous section, **According to Hoyle**.

Gravitational Waves

Gravity waves are believed to be ripples in the fabric of space-time caused by cataclysmic events in the Universe. Plausible sources of GWs involve the rapid accretion of large clumps of matter onto a neutron star or black hole. The largest such source is predicted to occur near a central massive black hole where the gravitational waveforms appear to be much more complex.[166] The tidal forces at the event horizon produce the gravitational radiation as they tear apart the star before it accretes onto the black hole. Interestingly, the tidal forces at the event horizon are much smaller for a massive black hole than for a small black hole. An infalling star of 1 M_{Sun} will not be tidally disrupted by a black hole of mass greater than 10^8 M_{Sun}. In this case the mass of the black hole grows through accretion, emitting some smaller amount of gravitational radiation and no significant electromagnetic radiation.

The effect of GWs on electromagnetic radiation is at most negligible. Yakov Zel'dovich (1973) made this clear when he showed that the effect disappears for all practical purposes in the presence of atoms and free charges.[167] Large GWs are associated with large gravitational wells, however, and they thereby impart important information for the study of gravitational redshift. Sigl *et al.* (2007) considered a model of massive stars that form in the outer regions of the accretion disk, collapse into compact objects, and inspiral toward the black hole.[168] Nearly two dozen assumptions were made in order to derive the resultant GW spectra: the rate of inspirals is approximately a few hundred per year; the fraction of accreted matter in the form of compact objects is one percent; the accretion efficiency of converting compact objects to GW radiation is twenty percent; compact objects inspiral one at a time in circular equatorial orbits; *etc.*

Figure 4.12 reveals the calculated characteristic GW amplitudes*, h_c, from compact objects with mass of 10 M_{Sun} inspiraling toward BHs with masses of 10^8 M_{Sun} (yellow), 10^7 M_{Sun} (red) and 10^6 M_{Sun} (blue), respectively. The solid and dashed lines represent black hole spin rates of 0 and 0.95, respectively. Note the sharp drop in amplitude in all cases beyond a critical frequency. Note also that the values for h_c never exceed 10^{-18}. By comparison the typical amplitude of (100Hz) gravitational waves emitted by bi-

---

* Characteristic amplitude, $h_c = \Delta L/L$, where $L = L_x = L_y$ is a specified interferometer length, and $\Delta L(t) = \delta L_x - \delta L_y$.

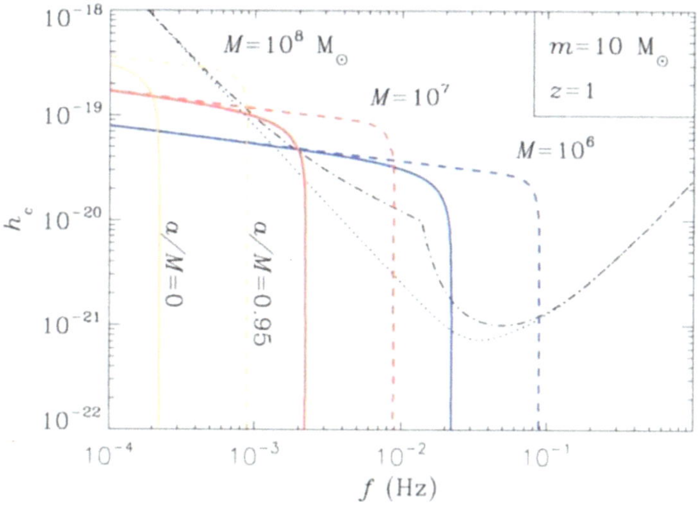

Figure 4.12 – Characteristic GW amplitudes for individual inspiral events
© 2007 American Physical Society

nary systems in the Virgo cluster of galaxies (~20 Mpc distant)* has the value of $h_c \sim 10^{-21}$.[169] Such small amplitudes of *typical* black hole accretion do not register any significant change in redshift distance measurements. More significant GWs will occur from binary systems composed of neutron stars and black holes. The most violent binary event is the black hole *merger*.

Baker *et al.* (2007) used numerical simulations to develop the GW waveform of a merging equal-mass, non-spinning black hole binary in circular orbiting trajectories.[170] The late-inspiral waveform shown in Figure 4.13 reveals the increasing strength (and frequency) of the radiation over simulated time frames. After the black hole event horizons have merged at time $t = 0$, the system goes into its final phase, merger-ring-down, and the radiation drops toward zero from the peak amplitude of $h_c \sim 0.25$.

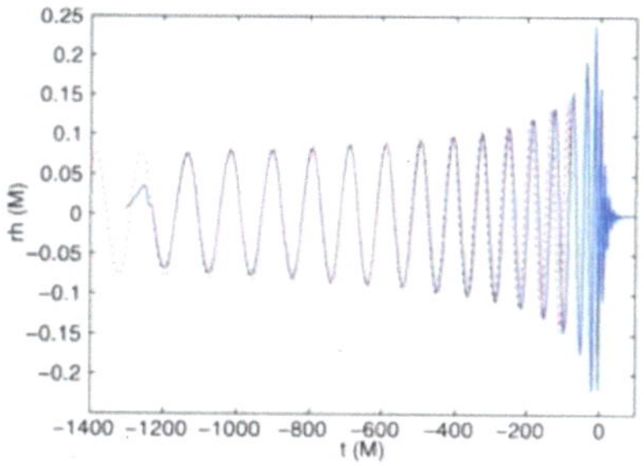

Figure 4.13 – GW waveform of BH merger
© 2007 American Physical Society

* This is the largest distance a first generation ground-based GW interferometer can probe.

The large dosage of gravitational radiation during a merger could be the signature of a very large gravitational well. Large GW interferometers could serve as a divining rod in the search for any heretofore undiscovered source of gravity in the galaxy. Gravitational wave strength cannot be considered another variable redshift factor *per se*, but it may provide support to any hypothesis that can predict its presence in abundance.

Dark Matter

The effect on gravitational potential of a spinning supermassive black hole is non-negligible. Distance measurements based upon redshift thereby entail some degree of uncertainty. However, the measurement errors from these factors cannot be so great as to call into question the fundamental results of astrophysical research. Quasars at $z > 3$ must still be at least ten billion light-years away unless $\mathcal{N}$-body gravitation or other larger factors are at play. Fred Hoyle's models of an extremely dense nuclear star cluster and magnetic monopole annihilation are two such factors that would abide the $z > 3$ quasars at much smaller distances from Earth. Another possible major factor is dark matter.

It is widely believed that most of the matter in the Universe is in some exotic form outside the standard model of particle physics. It does not interact electromagnetically, it is *dark matter*. It *does* respond to gravitational interaction. Indirect detection of dark matter is possible, as its mass acts as a source of gravitational attraction on luminous matter. Until the 1970s, the notion of dark matter was highly speculative. That changed when Vera Rubin started plotting rotation curves for spiral galaxies. By measuring the movements of ionized hydrogen gas clouds inside the galaxy, she discovered that orbital velocities held constant with increasing distance from the galactic center. Rubin *et al.* (1980) documented the rotation curves for twenty-one Sc-type galaxies that cover a range in luminosity from 3×10^9 to 2×10^{11} L_{Sun}, mass from 10^{10} to 2×10^{12} M_{Sun}, and radius from 4 to 122 kpc.[171] Figure 4.14 plots a superposition of all the rotation curves. After a rapid rise in velocity to ~125 km s^{-1} at the distance of ~5 kpc, rotation remains nearly constant to the farthest measured point. The cumulative mass must continue to increase with radial distance as the cumulative amount of light levels off. None of the galaxies exhibit falling rotation curves as required by Newtonian mechanics. Battaglia *et al.* (2005) used

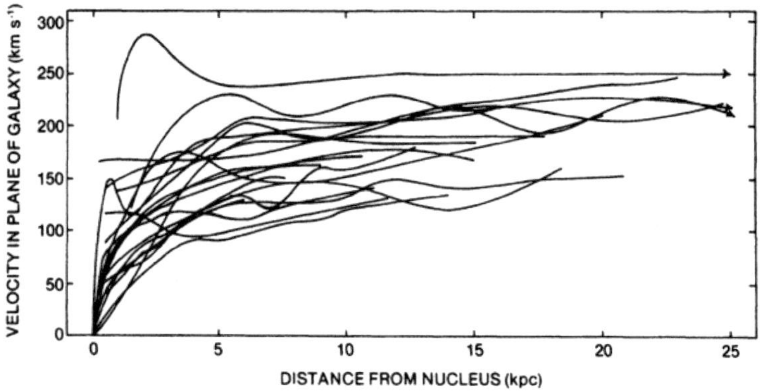

Fig. 4.14 – Rotation curves of 21 spiral galaxies

field blue horizontal branch stars, satellite galaxies, red giant stars, and globular clusters as additional tracers to measure the radial velocity dispersion profile of the Milky Way. Their compilation showed an almost constant value of 120 km s^{-1} out to 30 kpc and then a continuous decline to 50 km s^{-1} at about 120 kpc." [172]

The inescapable conclusion is that significant non-luminous matter exists beyond the optical galaxy image. The additional mass seems bizarre in light of the well-verified Tully-Fisher (TF) relation: $M \sim -10 \log V_c$, where V_c is the galaxy circular velocity. The TF equation explicitly ignores any effect of dark matter on the gravitational potential. This presents the paradox that rotation velocities of disk galaxies are scaled by their luminosities, even though dominated by dark matter. Such strange coupling between the amounts of dark and luminous galactic mass must offer one of the very few clues into the nature of the enigmatic dark matter.

Many space and ground-based observatories have been constructed in the attempt to discover the nature of dark matter. [173] While the density distribution has become better understood, its substantive nature remains elusive. We now know that the main baryonic mass content of galaxy clusters is in the intracluster medium (ICM), not in the galaxies themselves, and that dark matter constitutes ~80% of the total cluster mass. [174] Over a scale length of 1 Mpc, 95% of the measured mass is dark. We also know that many ordinary elliptical galaxies are highly deficient in dark matter. [175] Strigari (2013) provided an exhaustive report on all galactic searches for dark matter. The results firmly established that dark matter does not clump into objects as massive as dwarf stars or planets, thus indicating that the dominant matter component is more diffusively distributed.[176] This has led to weakly interacting massive particles (WIMPs) as the favored theory for the substance of dark matter. But searches for WIMPs have also led nowhere. Strigari suggests, "If WIMPs are not detected within the next decade, it is likely that there will be a paradigm shift in how dark matter is viewed in astrophysics, cosmology, and particle physics. Observations that are sensitive to deviations from the standard CDM model will become more relevant."

Another important aspect of *dark* observations is the search for any possible dark matter self-interaction. A dark force becomes necessary to govern the dark matter interaction, and that dark force requires dark boson particles. The possibility of having a nonzero cross section* has far-reaching astrophysical implications. The standard model predicts anisotropic scattering with a cross section $\sigma_{DM}/m = 0.6$ cm^2/g, similar to nuclear cross sections. The ongoing merging of galaxy clusters seems an efficient way to separate dark from baryonic matter and thereby estimate the interaction of the dark matter. Galaxy clusters contain gas clouds and dark matter, both of which are far more massive than the stars. Stars are largely unaffected by cluster collisions, but the intergalactic gas clouds suffer ram pressure from gravitational and electromagnetic forces to form a large central mass observable in the X-ray spectrum. Clowe *et al.* (2006) provided gravitational lensing maps of the Bullet Cluster collision that trace the gravitational potential *vs.* the X-ray plasma.[177]

Figure 4.15 is a color image of 1E 0657-558, the Bullet Cluster ($z = 0.296$). X-rays (imaged in yellow-red colors) depict the gas clouds, and the green contours trace the gravitational potential. The two different galaxies have moved ahead of the plasma after the collision. The gravitational potential does not trace the dominant baryonic component of gas clouds, it traces something much more massive in the vicinity of the two galaxy star concentrations. This is further direct empirical verification of the existence of dark matter. Markevitch *et al.* (2004) analyzed the Bullet Cluster for various constraints on the dark matter interaction cross section. The test for drag yielded $\sigma_{DM}/m < 1.25$ cm^2/g. A more sensitive constraint of

* The cross section measures the amount of interaction between an incident particle and a target particle of the same species. The larger the dark matter cross section, the more likely two dark matter particles will interact.

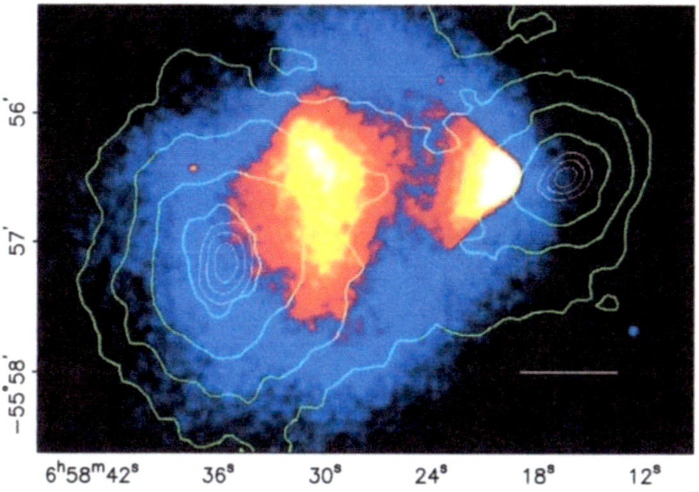

Figure 4.15 – Bullet Cluster collision

$\sigma_{DM}/m < 1.0\ cm^2/g$ comes from the consistency of the subcluster mass-to-light ratio with the main cluster (which rules out a significant mass loss due to the dark matter).[178] Any amount of dark matter interaction during a cluster collision will drag the particles toward the trailing edge of the cluster. The scatter analysis of 72 observed galaxy collisions by Harvey *et al.* (2015, 2017)[179][180] shows that the position of the dark mass remains within ~5.8 ± 8.2 kpc of associated stars, further constraining the self-interaction cross section to $\sigma_{DM}/m < 0.47\ cm^2/g$. Definite verification of dark matter interaction remains elusive.

Beyond its very existence, fifty years of observational research has added relatively little to the understanding of dark matter. Table 4.4 is an abbreviated list of the observational findings.

DM density	$M/L \sim 30h\ (M/L)_{Sun}$ at a radius of ~40 kpc [*] $M/L \sim 350h\ (M/L)_{Sun}$ at a radius of ~1 Mpc
DM interaction force	$0 < \sigma/m < 0.47\ cm^2/g$
DM substance	Unknown
DM halo shape	Conflicting opinion
DM clustering factor	Unknown
DM angular momentum	Unknown

Table 4.4 – DM properties based on observation

[*] The dimensionless h is a cosmological artifact defined as $h = H_0/(100\ km/s/Mpc)$. Currently, $h \sim 0.75 \pm 0.05$.

During this same time, however, $\mathcal{N}$-body simulations have combined the limited knowledge of DM with other observational constraints to compose an elaborate picture of the dark fabric of the Universe and how it evolved from small perturbations in the early cosmic density. Spiral galaxy rotation curves indicate that the galaxies are embedded in dark halos that are extended well beyond the galaxies themselves. This suggests a theory whereby dark halos form through hierarchical clustering and produce potential wells that subsequently condense the gas to form luminous content. The introduction of *cold* dark matter overcomes the issues of thermal and radiative pressures that otherwise prevent ordinary baryonic matter from collapse in the early universe. The CDM model is constrained by many other independent observations: CMB anisotropy, light-element abundances, distance-luminosity relation of Type Ia supernovae, the number density of clusters, cosmic shear due to gravitational weak lensing, large-scale structure, and the presence of the Lyα forest.[181] Numerical simulations of the well-constrained CDM halo model may thus provide possible answers to DM unknowns.

In the first simulation performed with CDM initial conditions, Navarro and White (1994) found that gas settles rapidly into centrifugally supported discs at the centers of small dark matter clumps.[182] Researchers then conducted numerical simulations of galaxy and halo formations with increasing numerical resolution. For example, Diemand *et al.* (2007) simulated the co-evolution of a parent halo and its subhalo population with 234 million particles on NASA's Project Columbia supercomputer. The run was completed in 320,000 CPU hours! Table 4.5 is an abbreviated list of findings from some of the most well known dark halo simulations.

Simulations show that dark matter concentrates over time in dense clusters. A measurable effect on redshift would occur if a mere 1% of the dark matter in a galaxy were to concentrate inside the nuclear cluster. This never happens for a normal galaxy. The situation is no longer normal, however, when galaxies merge. The most drastic changes occur during a major merger by the process of *violent relaxation*.

When two near-equal-mass galaxies with small relative velocities pass through each other, a major merger occurs. Major merging is most common in high-density environments.[188] Both galaxies are enclosed in large halos of dark matter. When the halos collide, the dynamical friction causes

After falling onto massive halos, subhalos survive for a long period orbiting within the potential well of its host.	Klypin *et al.* (1999) [183]
Tides remove subhalo mass from the outside in, which leads to strongly increasing subhalo concentrations toward the host center.	Diemand *et al.* (2007) [184]
The radial distribution of subhalos within their parent halos is substantially less concentrated than the dark matter as a whole.	Gao *et al.* (2004) [185]
Dark matter particle clustering is extreme (highest resolution shows ⅓ of particles collect in a region 10^{-8} of simulated volume).	Springel *et al.* (2008) [186]
Violent mergers of dark matter halos dominate the early phase of galaxy halo formation.	Mo and Mao (2004) [187]

Table 4.5 – DM properties based on simulations

the galaxies to decelerate, spiral ever closer to each other, and eventually merge. Hester and Tasitsiomi (2010) used simulated data to calculate a period of ~200 Myr for the final galaxy merger in a 1:1 halo merger.[189] Stars rarely collide during the merger, but gases do. When gases at different velocities collide, they are shocked, lose energy, and move deeper into the gravitational well where they should provide the fuel for active galactic nuclei.[190] The shocks can cause an entire galactic disc to flip and counter-flip over a period of several 10^8 years.[191] Major mergers of galaxies are also expected to be accompanied by large and rapid changes in the gravitational potential, $\phi(\mathbf{x},t)$, of the system. The return to equilibrium from the time-varying $\phi(\mathbf{x},t)$ is known as violent relaxation. The effect of violent relaxation is very strong and has significant effect on the differential energy distribution of the remnants. The deepening gravitational potential provides new energy states in which both dark and luminous particles are scattered. Since more dark matter than luminous particles are scattered into the highly bound regions, the end result is a higher dark matter fraction at the center of the final system.

Hilz *et al.* (2012) illustrated the temporal evolution of the energy distribution of bound particles through successive phases of violent relaxation, (see Figure 4.16), where ε represents the energy per unit mass, σ_t represents the Gaussian width at time interval t, and $N(\varepsilon)$ is the number of particles at corresponding energy ε.[192]

Colored curves circumscribe the particles initially within the energy range $-0.85 < \varepsilon < -0.81$. The truncated black curve indicates the range of all particles before merger. Simulations by Hilz *et al.* make clear the fact that

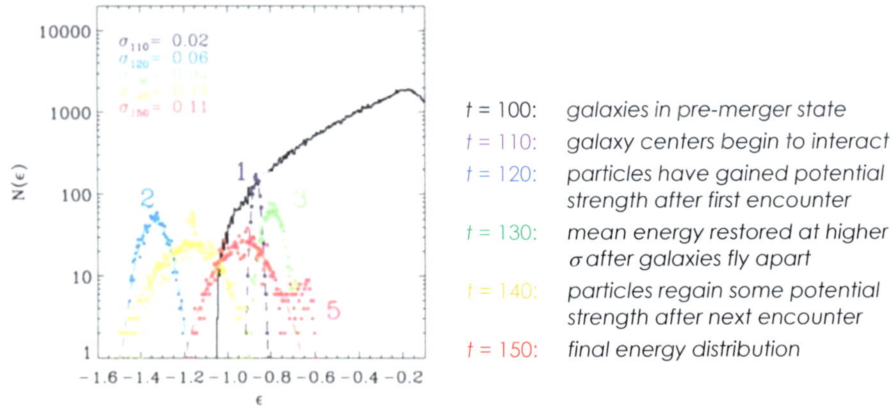

Figure 4.16 – Temporal energy distribution of near-equal-mass merger

the deepest gravitational potential well occurs after the first encounter of a near-equal-mass merger and that its maximum strength is not sustainable.

Observations and high resolution $\mathcal{N}$-body simulations have provided many pieces to the dark matter puzzle. Yet the picture remains blurred. There remain many important unsolved problems, and it is still unclear whether the halo shapes inferred from observations are consistent with the predictions of CDM cosmology.[193] This is no surprise given the high accuracy required for the $\mathcal{N}$-body simulation (input) parameters: luminosity, overall mass ratio, relaxation timescales, and redshift. Furthermore, major dark halo mergers often correspond to substantially smaller stellar mass ratios, thereby making the predicted luminous mergers unobservable.[194] [195] A surprising contradiction to the current theory of halo/galaxy mergers came recently from the discovery of the system known as Was 49. The dwarf galaxy (Was 49b) in this merger hosts a luminous AGN with an active SMBH hundreds of times more massive than expected, while the larger galaxy (Was 49a) has a black hole that is relatively quiet.[196] Dark halo theory is still in its infancy, and more surprises seem inevitable in the near future.

It is not possible to make any accurate estimate of the dark matter factor of gravitational redshift. Some $\mathcal{N}$-body simulations have shown that ⅓ of the particles collect in a region 10^{-8} of the simulated volume. That would be sufficient to cause some measurable effect on redshift. The large redshift proposed by Hoyle's original model would require a volume several orders of magnitude more dense in cases of galaxies similar to the Milky Way. Massive DM halos and/or $\mathcal{N}$-body gravitation of collapsed relativistic stars would ease this density constraint. Furthermore, subhalos cluster toward the center of their parent halo. The collection of potential wells would further increase the redshift in the nested fashion described by Hoyle's later model of magnetic monopole annihilation. It is quite possible that future DM/$\mathcal{N}$-body simulations will disclose a significant gravitational influence on electromagnetic radiation. Chapter 3 has revealed many AGN phenomena without adequate explanations. Given the infancy of DM research, given the computational burden of many-body simulations, and given the long list of unusual observations unexplained by CH, a variable AGN redshift hypothesis based upon underestimated factors of dark matter and $\mathcal{N}$-body gravitation seems at least *plausible*.

EGADS – A Hypothesis of AGN Variable Redshift

Extreme Gravity from Aggregated Dark Subhalos (EGADS)

The time varying gravitational potential produced by the merger of two equal-mass dark halos will cause a variable redshift in the electromagnetic radiation from the associated galaxies. The gravity becomes extreme at the center of the potential well where the dark subhalos aggregate. The infall of shocked gas into the well can cause one or more galaxies to go into an active state. A gravitational redshift factor of the active galactic nucleus will range from zero to a maximum value when the galaxies reach their closest point of interaction.

EGADS provides a plausible alternative to CH for extremely redshifted galaxies. It has rational explanations for problems that stupefy CH. Both EGADS and CH must stand or fall on the basis of their ability to explain all observations. Chapters 5 and 6 discuss recent observations and data analysis which help support the EGADS hypothesis.

Summary

Precision measurement of AGN redshift is frustrated by many factors. Proximity to the black hole, black hole spin, and relativistic galactic bodies are all gravitational factors that can render redshift measurement imprecise. Dark matter is a gravitational factor whose impact on redshift may be anything from nil to extreme. Fred Hoyle offered two different theories to explain the large redshift of the quasars. The substance of his pioneering theories serves as a guide to a new theory, EGADS, a plausible alternative to the cosmological hypothesis. EGADS hypothesizes that dark subhalos aggregate in a deep gravitational well at the center of the parent halo. An extreme, albeit temporary, gravitational effect occurs during a dark halo merger.

EGADS does not formulate any new physics. It stands in contrast to the favored cosmological hypothesis to explain the mysterious nature of high-z AGNs. Both hypotheses are open to doubt; both must rest on the totality of all observations.

Chapter 5

Recent Observational Evidence

Many peculiar phenomena are associated with active galactic nuclei. Chapter 3 discusses some of the odd issues that remain without good explanations. The issues mostly vanish with the simple explanation that *some* quasars are located at a distance less than prescribed by cosmological redshift. Such an assertion is easy to make, but it is extremely difficult to validate. EGADS is a *Gedankenexperiment* that assumes some quasars are less distant by virtue of their predicted variable gravitational redshifts. While no performable experiment exists for EGADS, the theory does make predictions that recent observational evidence may support.

EGADS Theoretical Predictions

1. *Luminous matter from galaxies falls into a common potential well.* – Intergalactic matter will experience high gravitational potential during the dark halo merger. Luminous matter will appear directed toward a common point.
2. *Black hole mass increases with redshift; low-mass holes become scarce at high z.* – As the gravitational factor increases with *z*, the calculated black hole mass becomes increasingly overestimated.
3. *Black hole accretion rate increases with redshift.* – Large amounts of matter are released down the gravitational well due to the angular momentum lost during the merger.
4. *An AGN becomes 'invisible' during the merging event.* – Galactic dust will enshroud the galaxy and suppress optical/UV emissions.
5. *High ionization lines will blueshift relative to Hβ emission.* – The merger induces a strong magnetic field that affects atomic emissions and alters the direction of inbound gases.

6. *Lyman-alpha forests will display strong sightline-to-sightline variance.* – Mistaken quasar distances will show a scatter in the median optical depths of H_I Lyα (λ_{rest} = 1216 Å) and He_{II} Lyα (λ_{rest} = 304 Å).

Luminous Bridges

The standard model for active galactic nuclei does not provide a good understanding for how far-away particles in the galaxy can lose their angular momentum and approach the central SMBH. The favored explanation involves mergers or other external interactions. This is observed in numerous AGNs that appear to have experienced recent merger activity. Figure 5.1 is an image from Villar-Martin *et al.* (2010) that reveals ionized [O_{III}] gas extending from the nucleus of Seyfert galaxy SDSS J0123+00 (on the right) to the galaxy G1, a distance ~133 kpc.[197] The gas nebulae strongly suggest that the origin is tidal debris from an encounter with the galaxy. Luminous debris that forms a tidal bridge is well suited for detailed observations and offers the best opportunity to evaluate EGADS prediction #1. Although the number of optical associations of galaxies with very different redshifts has steadily increased through the years, most of the astronomical community has not made any effort to study the phenomenology. When Martin Lopez-Corredoira and Cristian Gutierrez (2006) were given observatory time for such study, they found many of the objects contained yet more anomalous characteristics than previously documented. Their findings include the study of NGC 7603, NEQ3 and MCG 7-25-46, all of which have luminous bridges that apparently interconnect objects with different redshifts.[198]

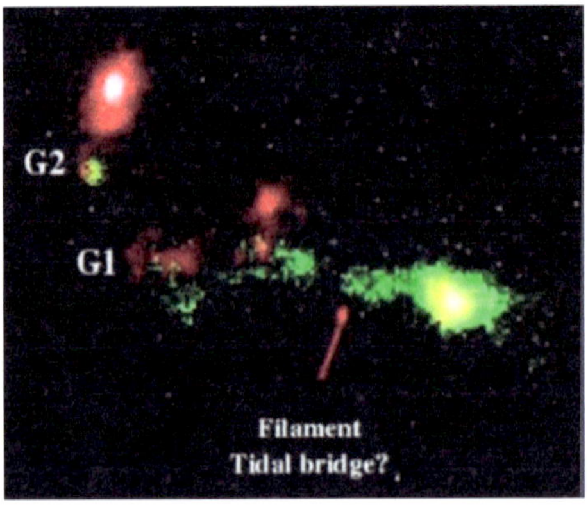

Figure 5.1 – SDSS J0123+00
Green: Continuum subtracted [O_{III}] λ5007
Red: Normal continuum in regions of interest
© 2010 Oxford University Press

Among the more interesting bridges is the one found in system NGC 7603. Figure 3.16 reveals the bridge and the disparate redshifts of the four-object system. The configuration holds special interest because NGC 7603 and NGC 7603B are separated by less than one arcminute in the sky and the two additional objects are contained within the luminous bridge. In Figure 5.2 the smaller, higher-z objects (encircled) can be seen to have deformed tails, suggesting further interaction. Spectra of the smaller objects show line ratios similar to H_{II} galaxies with very intense Hα (extended width of ~80 Å and ~160 Å) indicative of vigorous star formation. Spectra of the luminous bridge indicate an estimated redshift of $z = 0.030$ (0.001 greater than its source galaxy NGC 7603).* The observers could not gain enough information to build a 3-D representation, but it only requires a little imagination to picture an EGADS environment. Everything is consistent with luminous material from both NGC 7603 and NGC 7603B falling in a strong gravitational field toward the 'more distant' objects. In three dimensions the bridge actually may narrow and extend beyond the smaller galaxies.

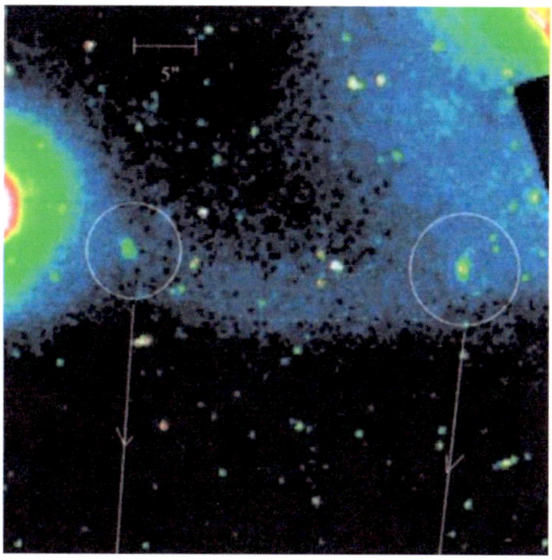

Figure 5.2 – system NGC 7603
© 2006 AIP Publishing

Another intriguing four-object system is NEQ3 depicted in Figure 5.3. The foreground object is a lenticular (S0/Sa) galaxy at redshift $z = 0.1239$. The three disturbed background galaxies are at redshifts $z = 0.1935$, 0.1939 and 0.2229, all separated by less than ~3 arcseconds across the sky. The redshift of the filament (bridge) is identifiable by the cross-correlation function with a maximum $z = 0.19$ and a secondary maximum $z = 0.12$.

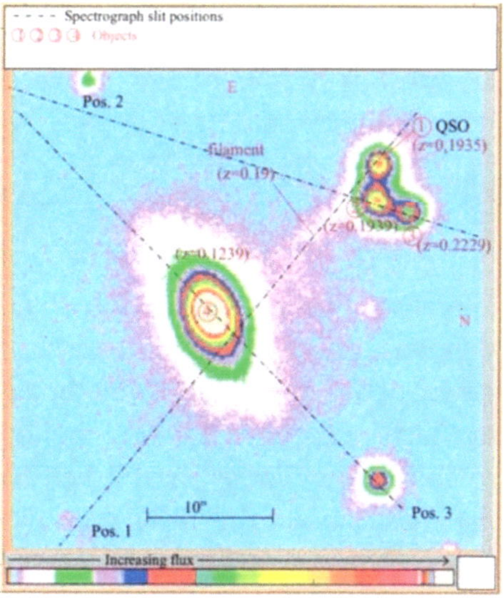

Figure 5.3 – system NEQ3
© 2006 AIP Publishing

* A bridge with redshift *less than* the source will form *behind* the dark merger shroud where it is hidden from view.

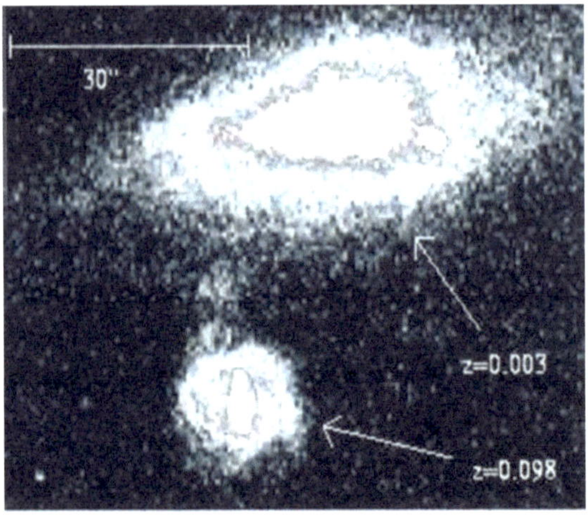

Figure 5.4 – system MCG 7-25-46
© 2006 AIP Publishing

Again it is easy to imagine the luminous material falling toward a common point in the vicinity of the background galaxies.

Figure 5.4 shows the system MCG 7-25-46 with a filament interconnecting two objects with redshifts of $z = 0.003$ and $z = 0.098$. Hα emission of the filament at $z = 0.003$ abuts the observed emission at $z = 0.098$. This is a most strange coincidence. Is it not consistent with matter that is falling down a gravitational well?

The recent observations of Lopez-Corredoira and Guiterrez must whet the appetite of any serious astrophysicist. Disparate redshifts of interconnected objects disfavor CH. Disrupted objects with filaments $z_{obj1} < z_{filament} < z_{obj2}$ favor a gravitational well. If the CH model is to remain unscathed, there must be further observations of the many documented anomalous redshifts.[108][199] Closer examination of the bridge between NGC 4319 and Markarian 205 (seen in Figure 3.15) would be a fine place to begin.

Black Hole Mass Estimations

The black hole mass is a fundamental parameter in any description of active galactic nuclei. Good estimation of the masses across cosmic epochs is essential to AGN science. In the last two decades estimation methods* have improved considerably whereby each successive method builds upon the former in a manner akin to the *distance ladder* established to measure the Hubble constant.† The first step came in the 1990s with the discovery that a galaxy's central black hole mass is closely related to its surrounding spheroidal distribution of stars. The central black hole mass is

* Peterson (2014) provides an excellent overview of all the methods of black hole mass measurement.[200]

† See Chapter 1, **The First Question: How Accurate is H_0?**

tightly correlated with the bulge mass, which in turn scales directly with the velocity dispersion, σ, of the stars in the host-galaxy bulge. The M_{BH}-σ relationship allows an estimate of the local black hole mass function by convolutions with galaxy bulge distribution functions. Measurement of M_{BH} in this manner is not possible without adequate stellar resolution.

The AGN model includes a broad-line region (BLR) of clouds surrounding the central engine at a radius of the order 10^{16}–10^{19} cm. Although the BLR remains spatially unresolved, observed Doppler-broadened emission lines are consistent with orbital speeds of 10^4 km/sec that clouds would attain at this distance from the black hole. With the assumption that the BLR is virialized and dominated by the gravity of the supermasive black hole, it is possible to estimate the mass of the central (black hole) source by application of the virial theorem:

$$M_{BH} \approx G^{-1} R_{BLR} v^2,$$

where G is the universal gravitational constant, R_{BLR} is the radius of the BLR, and v is the RMS velocity (line width) of the line-emitting gas. With the further assumption that the broadened spectral lines are powered by photoionization from the continuum radiation produced by the central engine, variations in the continuum will lead to changes in the line emission delay by the time $\tau = R_{BLR}/c$. Substitution then yields

$$M_{BH} = G^{-1}(c\tau) f v^2,$$

where f is a scaling factor that depends on the geometrical distribution of the BLR gas. In practice the value of f is empirically determined by assuming that the derived masses are consistent with those predicted from the M_{BH}-σ relation of local inactive galaxies. The inferred BLR scales are light-days to light-months for low luminosity AGN and light-years for high luminosity quasars. Measurement of the lag between continuum and the emission-line variations is known as reverberation mapping (RM), where $c\tau$ is considered the reverberation radius. Changeover from angular resolution to time resolution makes mass measurement possible for the more distant quasars.* McLure and Dunlop (2004) offered several types of evidence that demonstrate that dynamics of the BLR appear to be consistent with the virial assumption.[201] For example, the virial relationship is found

* Measurement of R_{BLR} by reverberation mapping is only possible for (type-1) AGNs that exhibit broad line regions.

consistent with M_{BH}-σ for every AGN for which emission-line lags and line widths have been measured. The difficulty with RM is that it is an extremely time consuming method that requires a minimum of 30-50 observations over several years in order to acquire one reliable measurement. As a consequence RM has only been used to estimate the virial mass in a very small number of objects. It is clear that something further is required for large-scale black hole mass estimations.

RM surveys revealed a crucial relationship between the size of the broad emission-line region and the AGN luminosity. This R_{BLR}-L relationship produces the updated equation

$$M_{BH} = G^{-1}\left(\lambda L_{\lambda}\right)^{\alpha} f v^{2},$$

where $(\lambda L_{\lambda})^{\alpha}$ is the parameterized substitution for R_{BLR} at the selected emission line wavelength λ, and $\alpha \sim 0.5-0.7$. Corbett *et al.* (2003) used composite spectra from more than 22,000 QSOs to find more than 99% significance in the correlation between velocity width and continuum luminosity for Hβ, Hγ, C_{III}], Mg_{II} and C_{IV}.[202] A clear and reliable correlation is shown between the radius of the Hβ emitting region and the monochromatic luminosity at 5100Å. The $R_{BLR(H\beta)}$-L_{5100} relationship can be used to estimate R_{BLR} in every source of known luminosity and to thus obtain the BH mass by combining this size with the observed Hβ linewidth. This measurement method is known as the *single-epoch mass estimate*, and the lines used in single-epoch mass estimation are known as *single-epoch estimators*. The Hβ estimator is the most reliable, but it is limited by the fact that Hβ is redshifted out of the optical at $z > 0.8$. This problem may be overcome by using Mg_{II} and C_{IV} as rest-frame ultraviolet proxies for Hβ. Using simultaneous UV, optical and infrared spectroscopy on an AGN sample at $z \sim 1.55$, Mejia-Restrepo *et al.* (2016) recalibrated M_{BH} based on Mg_{II} and C_{IV} estimators.[203] The Mg_{II} estimator gives results consistent with Hβ, but C_{IV} shows significant evolution with redshift.

Researchers were quick to exploit the UV black hole mass estimators to explore the evolution of the quasar mass-luminosity relation. Corbett *et al.* (2003) calculated a relationship of $M_{BH} \sim L^{0.97\pm0.16}$. They found that the lack of significant evolution in the velocity-luminosity correlation places strong constraints on the evolution of black hole masses for a fixed luminosity: $M_{BH} \sim (1+z)^{\beta}$, $\beta < 1$. This is smaller than necessary to explain the evolution of the quasar luminosity function with a single population of

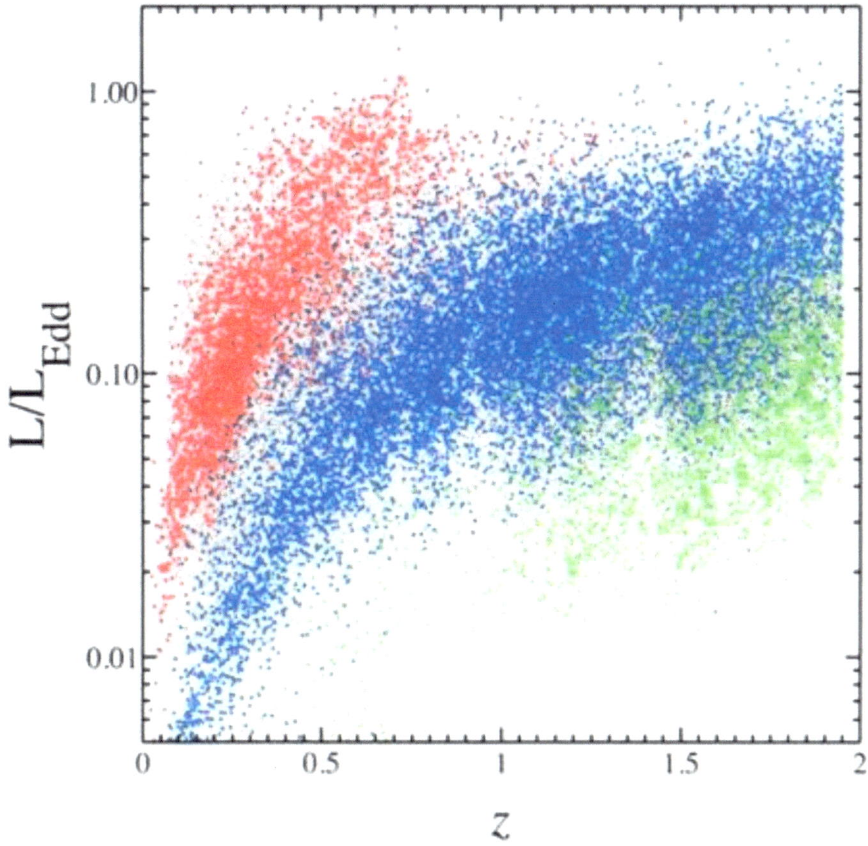

Figure 5.5 – L/L_{Edd} vs. redshift for entire SDSS type-1 AGN sample
Colors highlight subsamples in the different mass ranges
Red: $M_{BH} \sim 10^7\ M_{Sun}$
Blue: $M_{BH} \sim 10^8\ M_{Sun}$
Green: $M_{BH} \sim 10^9\ M_{Sun}$
© 2012 Oxford University Press

long-lived objects.[202] Trakhtenbrot and Netzer (2012) (hereinafter TN12) later examined the FWHM of the $Mg_{II}\lambda2798$ and $C_{IV}\lambda1549$ lines by observations on sources that also contain $H\beta\lambda4861$. They found C_{IV} cannot be used to obtain precise estimates of M_{BH} (for reasons unknown), but they did find $Mg_{II} \sim H\beta$ for all sources with $FWHM(Mg_{II}) < 6000$ km s^{-1}. The combination of single-epoch estimators $R_{BLR(MgII)}$-L_{3000} and $R_{BLR(H\beta)}$-L_{5100} were then used to obtain precise M_{BH} estimates for all quasars at $z < 2$. AGN spectra from the SDSS DR7 survey were analyzed with both estimators in order to frame a picture of the evolution of black hole mass (M_{BH}) and relative accretion rate (L/L_{Edd}) over the last ten billion years.* The observed sample comprises 20,894 ($z < 0.75$) Hβ objects, 43,995 ($0.75 < z < 2$) Mg_{II} objects, and 6,731 ($0.5 < z < 0.75$) objects in which both Hβ and Mg_{II} lines appear.† The final picture is reproduced in Figure 5.5.[204] Note the smooth transition at $z = 0.75$ where the Hβ and Mg_{II} estimates overlap.

The reader is encouraged to study Figure 5.5 closely. There is a clear trend toward more massive black holes with increased z. This appears to support EGADS prediction #2. L/L_{Edd} is larger for smaller mass black holes at all redshifts, although large-mass BHs at high z can accrete as fast as small-mass BHs at low z. This supports a previous study by Netzer and Trakhtenbrot.‡ Each color band shows a rise in accretion rate with increased z. This appears to support EGADS prediction #3. The steep rise of L/L_{Edd} flattens toward the value of 1.0, and the vast majority of very massive black holes ($M_{BH} > 10^9\ M_{Sun}$) do not show accretion close to 1.0 at $z \sim 2$, the epoch of peak AGN activity. This maximal value on accretion constrains the growth rate of the AGN population over the last ten billion years. Overall, this is a complex picture not easily explained by standard theory.

A concern of any astronomical survey is selection bias. Since M_{BH} and L/L_{Edd} are dependent on the source luminosity, both are subject to systematic errors when approaching the flux limit of the survey sample. When applied to large samples such as the SDSS, such errors tend to cancel, leading to presumptive statistical accuracy. The presumption seems to hold for single-epoch virial estimators because there is no systematic offset between them and the results of reverberation mapping. One underappreciated statistical bias that does lead to misestimating the average M_{BH} is the *Malmquist-type bias*. Mean BH mass becomes biased when variations of luminosity at some fixed *actual* BH mass scatter more 'lighter BHs' than 'heavier BHs' into a (statistical) luminosity bin. Mass is overestimated because the average luminosity value becomes higher than the *actual* luminosity value associated with the *actual* mass. Shen and Kelly (2010) calcu-

---

* See the next section for the definition of relative accretion rate, L/L_{Edd}.

† Samples from the 2QZ and 2SLAQ surveys were excluded due to lack of any additional information.

‡ Interestingly, Netzer and Trakhtenbrot (2007) also show the average BH growth rate for $z < 0.75$ AGNs seems to agree with the star formation rate over the same redshift interval.[205]

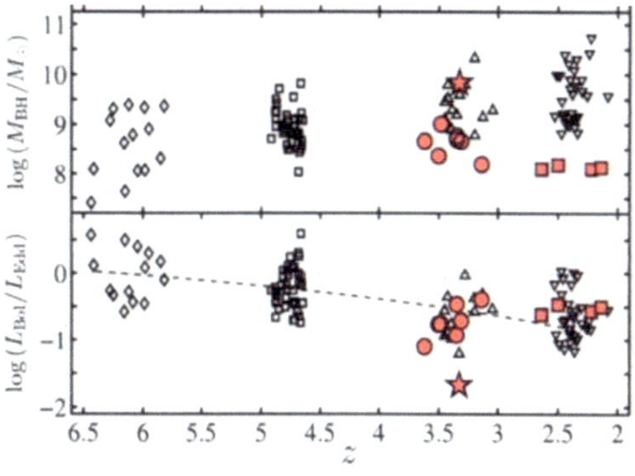

Figure 5.6 – M_{BH} and L/L_{Edd} at $z > 2$

lated that the mass bias contributes at least 0.2-0.3 dex* at $L_{bol} > 10^{46}$ erg s^{-1}. Given the Malmquist luminosity-dependent bias, the average virial BH masses in $z > 0.7$ SDSS quasars are likely overestimated by a factor of a few.[206] This does not fully explain the M_{BH} increase with redshift that is represented in Figure 5.5.

A better explanation would emerge from a picture extended beyond the boundary limit of $z = 2$. That image is difficult to develop. The necessary AGN samples must come from deeper, fainter objects that require multi-wavelength surveys with appropriate X-ray coverage. Several such recent surveys were able to image quasars an order of magnitude fainter with the use of K-band spectroscopy. Figure 5.6 represents the new data points (in red) alongside the older data (in black). The less luminous sources of the newer surveys have significantly lower BH masses, but there is an evident lack of lower accretion rates. This represents a possible confirmation at higher redshift of the picture (Figure 5.5) presented by TN12.† The red star represents an interesting object with a high mass of 10^{10} M_{Sun} and a low relative accretion rate of $L/L_{Edd} = 0.02$. The remarkable thing about Figure 5.6 is what is *not* seen. The density of fainter AGNs at $z = 3.3$ is expected twenty times greater than the more luminous population measured at $\Phi \sim 10^{-8}$ Mpc^{-3} in previous surveys. The specified flux limit should recover sources as low as $M_{BH} \sim 7 \times 10^7$ M_{Sun} or accretion rates as low as $L/L_{Edd} \sim 0.01$. Yet no lower mass objects were found.[207] Deeper X-ray surveys, such as Chandra COSMOS Legacy Survey, should detect a significant population of faint $z \sim 5$-6 AGNs with $L_{2\text{-}10\ keV} > 10^{43}$ erg s^{-1} and $M_{BH} = 10^6$ M_{Sun}, but this also is not observed.[208] This represents a most important *non-discovery*. If smaller black holes were scarce at higher *z*, the cosmological principle surely would falter. It is possible that the discrep-

* 0.3 dex is equivalent to 0.3 orders of magnitude, hence several times as large.

† It further confirms indirectly the earlier work of Hagai Netzer that showed the largest BHs reside almost exclusively in high-*z* quasars.[209]

ancy is due to some unknown evolution in radiative efficiencies and/or obscuration factors. It also may be validation of EGADS prediction #2.

Relative Accretion Rate

The theory of AGN evolution requires an accurate description of the frequency and rate at which matter is accreted. Astrophysical accretion is a complex subject made yet more difficult by the uncertain nature of black holes.* The subject at hand finesses much of the complexity with detailed definitions of *Eddington luminosity* and *relative accretion rate.*

Accretion onto massive objects and the associated release of the binding gravitational energy are important sources of radiation in astrophysics. The amount of energy released by the accretion of mass m onto the surface of a body of mass M and radius R is $\Delta E = GmM/R$, where efficiency is given by the ratio of M/R. In the case of active galactic nuclei, accretion energy may simply vanish into the black hole. Luminosity is given by

$$L = M'\eta c^2,$$

where M' (= dM/dt) is the infall rate of accretion (into the black hole), and η is the radiative efficiency introduced in Chapter 4.† If matter were to accrete infinitesimally slowly onto the black hole, η would approach unity and all rest mass energy would be extracted into radiation. For the accreting source to remain intact, the inward gravitational force must balance or exceed the outward radiation force. When the two forces are exactly balanced, a maximum luminosity is reached known as the *Eddington limit* or *Eddington luminosity*, L_{Edd}. The value $M'_{Edd} = L_{Edd}/\eta c^2$ is known as the *Eddington accretion rate.* The value of L_{Edd} represents the maximum luminosity for an object over some protracted steady-state accretion. For spherical accretion of fully ionized gas, $L_{Edd} \sim 1.5 \times 10^{38}(M_{BH}/M_{Sun})$ erg s^{-1}. The ratio $\xi = M'/M'_{Edd}$ is used to define a normalized scale-free relationship that involves the Eddington luminosity. $L/L_{Edd} \sim M'/M'_{Edd}$ is typically used in place of ξ. Since $L \sim M'$ and $L_{Edd} \sim M_{BH}$, the ratio L/L_{Edd} may be considered as the rate of accretion per unit of mass, otherwise known as the Eddington ratio or *relative accretion rate.* [211]

---

* See *Accretion Power in Astrophysics* for a thorough description of accretion onto compact galactic objects. [210]

† See Chapter 4, **Black Hole Rotation (Spin)**.

Redshift interval	Average redshift	Average QSO luminosity	Spectral line	FWHM (km s^{-1})	Log(λL/W)	Log($M_{BH}/M_\odot$)	Log(L/L_{Edd})
0.30, 0.68	0.556	−22.28	Mg_{II}	3546±19	37.379±0.001	7.8±0.33	−0.71±0.33
0.68, 0.92	0.803	−23.26	Mg_{II}	3875±14	37.747±0.017	8.1±0.33	−0.64±0.33
0.92, 1.13	1.028	−23.85	Mg_{II}	3878±15	37.977±0.030	8.3±0.33	−0.56±0.33
1.13, 1.32	1.224	−24.26	Mg_{II}	3783±16	38.127±0.040	8.4±0.33	−0.48±0.33
1.32, 1.50	1.414	−24.57	Mg_{II}	4104±18	38.229±0.049	8.5±0.33	−0.50±0.33
1.50, 1.66	1.552	−24.75	Mg_{II}	3889±31	38.272±0.055	8.5±0.33	−0.41±0.33
1.50, 1.66	1.585	−24.80	C_{IV}	5438±58	38.394±0.031	8.9±0.36	−0.79±0.36
1.66, 1.83	1.746	−25.03	C_{IV}	5444±44	38.472±0.024	8.9±0.36	−0.75±0.36
1.83, 2.02	1.919	−25.25	C_{IV}	5687±40	38.564±0.017	9.0±0.36	−0.75±0.36
2.02, 2.25	2.132	−25.46	C_{IV}	5629±39	38.609±0.010	9.0±0.36	−0.69±0.36
2.25, 2.90	2.445	−25.83	C_{IV}	5707±39	38.763±0.001	9.1±0.36	−0.64±0.36

Table 5.1 – Measured Line Widths of Mg_{II} and C_{IV}

Given that $L/L_{Edd} \sim L/M_{BH} = M'\eta c^2/G^{-1}R_{BLR}v^2$, accuracy of the relative accretion rate becomes dependent upon consistent radiative efficiency η and the given relationship of the emission line width v with black hole mass M_{BH}. These assumptions are fraught with danger. For example, McLure and Dunlop (2004) assumed FWHM is *approximately constant* to explain the evolution of L/L_{Edd} with redshift,[201] but Fine *et al.* (2006) showed FWHM evolution from composite spectra taken from more than 2000 QSOs at $0.5 < z < 2.5$. Table 5.1 reveals the trend found in Mg_{II} and C_{IV} line widths.[212] That leaves the L/L_{Edd} evolution without any reliable explanation.* EGADS prediction #3 states simply that M' increases with z_g, and hence z.

The most striking feature of Figure 5.5 is the well-defined array of color bands. It indicates that the smaller black holes are the faster accretors at any redshift. L/L_{Edd} would decrease for the more massive objects if they also contained wider FWHM. Such an explanation is possible for EGADS objects with higher z_g, but it is not plausible at low-z, where reverberation mapping has been substantiated and where z_g is negligible. Has selection bias once again distorted the picture?

Two publications of Yue Shen and Brandon Kelly are regularly cited for their use of a Bayesian inference model that overcomes the effect of the flux limit for the SDSS survey of approximately 58,000 quasars at $0.3 < z < 5$.[213] [214] The technique reveals demographics of broad-line quasars unattainable by observation alone. Their constraint of the black hole

* See the following section, **Accretion Disk Geometry**, for the cause of reduced radiative efficiency.

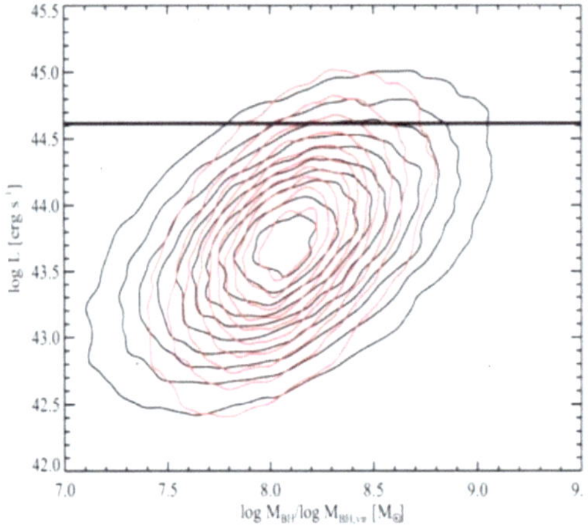

Figure 5.7 – *M/L* plane at z = 0.6

mass function (BHMF),* the luminosity function (LF), and the relative accretion rate throws added light on quasar evolution. The 'full forward' modeling is not without uncertainties, but it does bring new insights by accounting for both the flux limit and the difference between true and estimated BH mass. Among the interesting revelations is the lack of strong BH mass dependence on the *mean* Eddington ratio. This attribute is depicted most vividly by Shen (2013) and reproduced in the figures on this page.[215]

Figure 5.7 shows the mass/luminosity (*M/L*) plane for SDSS broad-line quasars at $z = 0.6$.† Red contour lines represent the *true* quasar distribution based on the Shen and Kelly model. Black contour lines represent the *observed* distribution when measured by Hβ single-epoch estimators. The *actual observed* distribution lies above the black horizontal line that delineates the flux limit of the sample. The combination of Malmquist-type bias and flux limit results in truncation of data, which overestimates mass and loses some population of lower accretors at fixed mass. Figure 5.8 illustrates the latter effect more vividly with color-coded contour lines. It includes two dotted diagonal lines to better estimate relative accretion

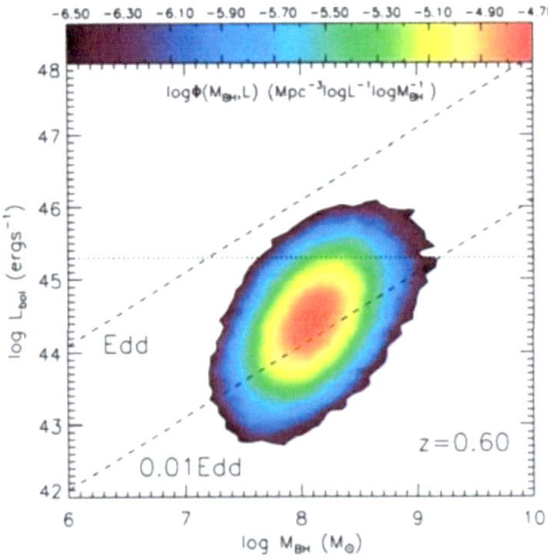

Figure 5.8 – *M/L* plane at z = 0.6

* Constraints on BHMF remain at a factor of a few due to inherent errors in the single-epoch mass estimators and severe flux limits on high-z SDSS quasars.

† Here *L* here is the monochromatic luminosity measured at restframe 2500 Å, where $L_{bol} \sim 5L$.

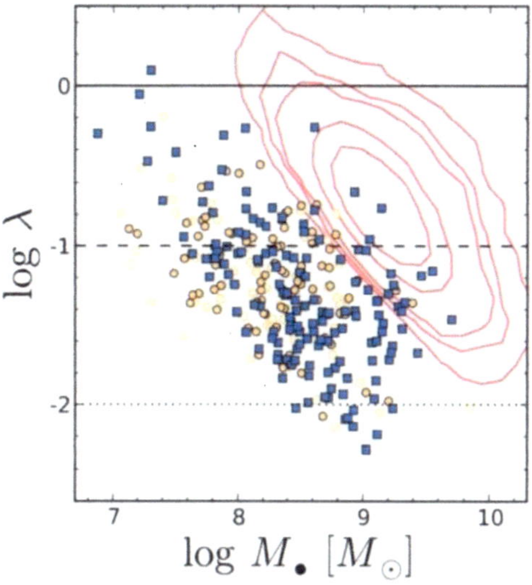

Figure 5.9 – L/L_{Edd} vs. M_{BH} distribution of observed SDSS, zCOS-MOS and VVDS samples
© 2015 Oxford University Press

rates. The mean Eddington ratio, $<L/L_{Edd}>$, is higher for flux-limited objects than the mean taken from the whole. Note especially how the flux-limited sample suggests a higher accretion rate for lower-mass objects, similar to Figure 5.5, but the mean rate changes little for different BH masses over the entire (forward modeling) sample. The technique of Shen and Kelly shows that a flux-limited sample will always preferentially select low-mass BHs at higher relative accretion rates. Netzer stresses that high-mass, high-accretion objects would appear in the distribution because they are not affected by the flux limit. Shen stresses that their non-appearance is illusory due to the lower number density of high-mass objects. The question becomes: what happens with an improved flux limit?

The deeper surveys of VVDS and zCOS-MOS ($1 < z < 2$) extend the range of SDSS by ~2 dex in luminosity, ~1 dex in M_{BH}, and ~0.5 dex in L/L_{Edd}. The supplementary objects observed by zCOS-MOS (blue squares) and VVDS (yellow circles) appear in Figure 5.9 with the same anticorrelation of L/L_{Edd} to M_{BH} as the SDSS observed objects (red contours). Note the continued absence of high-accretion, high-mass blue squares or yellow circles with the extended survey limits. Schulze *et al.* (2015) used these surveys to improve the reliability of Shen and Kelly with the use of a maximum likelihood-fitting model. They found a small (positive) correlation of L/L_{Edd} to M_{BH} at intermediate redshifts, but the anticorrelation reappears at low-z where massive BHs become inactive.* They also found evidence for an *increase of the average Eddington ratio with redshift*.[216]

The negative correlation of L/L_{Edd} and M_{BH} can be neither confirmed nor refuted. Recent observations would confirm, but samples remain flux-limited. Best models would refute, but their results are jeopardized by

* Schulze *et al.* (2015) used the term 'inactive' to refer to any AGN with $L/L_{Edd} < 1\%$.

misestimated factors. Furthermore, the bolometric luminosity becomes misestimated when the correct combination of black hole mass, spin and accretion rate is not taken into account. It now proves necessary to consider the accretion disk geometry.

Accretion Disk Geometry

The nature of the accretion disk exists in the realm of the theoretical, and it likely will remain so until greater optical resolution becomes possible. The inner radius of an accretion disk with black hole mass of $10^8\ M_{Sun}$ is less than 0.0001 pc (parsecs). The extreme outer dimension is of order 0.05 pc. The current smallest possible resolution of any galactic object is a few milliarcseconds corresponding to ~1 pc at a distance of 100 Mpc ($z \sim 0.03$). Hence, the only thing certain about accretion disk geometry is that it is not much larger than the solar system. A detailed picture of the accretion disk requires further knowledge of factors such as accretion rate, magnetic field strength, disk corona, source of the hard X-ray continuum, photon-pair production, and the functional form of the viscosity. All the uncertainty leaves open many possible alternative models for the accretion flow. Fortunately, the relative merit of the EGADS model is that it does not require an understanding of all the complexities built into modern accretion theory. It is only necessary to understand how different optical depths and different geometries impact the radiative efficiency η (see Table 4.3). The effective optical depth, τ_{eff}, depends on the column density and level of ionization of the gas: τ_{eff} becomes very large during periods of fast accretion, and photons become 'trapped' in the inflow. The geometry is indicated by the disk semi-thickness from the center, $H(R)$, where H is the vertical thickness (height) and R is the disk radial length from the center. A geometrically thin disk is defined by $H \ll R$. The geometrically *slim* and *thick* disks are defined variously either by $H \approx R$ or $H \gg R$.

Optically thick, geometrically thin disks have received most of the attention due to the fact that they are likely to form over a wide range of conditions in galactic nuclei during events of relatively heavy cold gas accretion. They are also much easier to treat both analytically and numerically than slim/thick models. While the thin model works well to provide a general overview of accretion disks, it fails in some important respects. The major difficulty is finding the mechanism by which distant particles in the galaxy lose most of their angular momentum.* Laor and Netzer (1989) found an upper limit of $L/L_{Edd} = 0.3$ for the thin disk.[217] This becomes untenable for any theory of AGN evolution that requires periods of super-Eddington accretion.† If the angular transport is cool, the thin disk becomes gravitationally unstable, breaks into lumps, and accretes even more slowly. If the angular transport is hot, the thin disk will blow away in a wind.[218] Stated more succinctly, thin disks do not have the effective turbulent viscosity to adequately channel the necessary 'fuel' into the tiny area of the galaxy occupied by the black hole. It becomes fundamentally necessary to expand the repertoire of accretion disk geometries.

Researchers have shown great interest over the last several decades in the development of theoretical accretion disk flows. It is here that theory and observation have come together rapidly to help explain the true nature of AGNs. The complexity of the subject has left a wide swath of theory to contemplate.‡ A survey of the literature reveals that there is no accepted standard model for a slim or thick accretion disk. Even their definitions are not standardized. The confusion level becomes reduced by recognition of two fundamental facts for the restructured geometries. First, their accretion flows are adiabatic. Gravitational energy, rather than become liberated through radiation, is locked up in the gas as thermal energy (or entropy) that may advect beyond the BH horizon or in an unbound outflow. The radiative efficiency η is thereby decreased substantially. Second, the rate of accretion can either diminish or increase by orders of magnitude. Any large divergence in accretion requires millions of years to transpire, however, and nothing should be expected over the career duration of the astronomer.[220]

Narayan and Quataert (2005) divided accretion flows into two major classes on the basis of their radiative efficiency η. For thin disks (vertical thickness typically in the range 0.1% to 3% of the radius), η is ~0.06-0.4. For thick disks (vertical thickness exceeds 20% of the radius), $\eta \ll 1$. Figure 5.10 shows their thin and thick classes represented by the clear and shaded

---

* The current favorite mechanism is magneto-rotational instability. Viscosity is produced by the turbulence from magneto-hydrodynamic (MHD) instabilities in a differentially rotating, magnetized disk.

† See Chapter 3, **Evolution**.

‡ See *Foundations of Black Hole Accretion Disk Theory* for a modern and thorough analysis.[219]

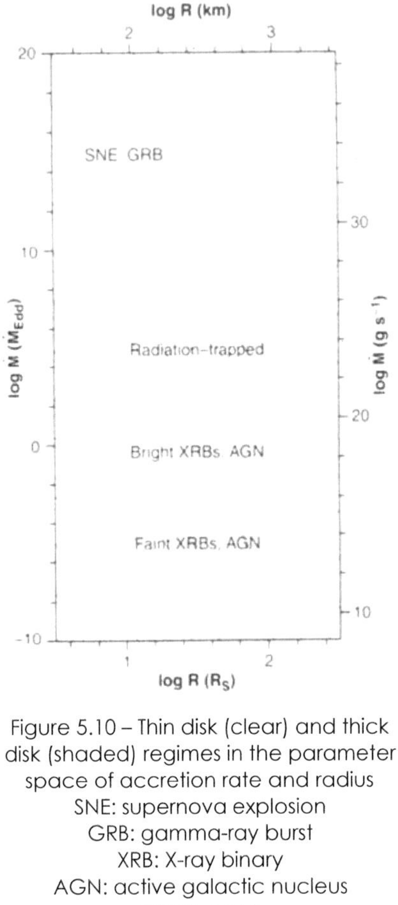

Figure 5.10 – Thin disk (clear) and thick disk (shaded) regimes in the parameter space of accretion rate and radius
SNE: supernova explosion
GRB: gamma-ray burst
XRB: X-ray binary
AGN: active galactic nucleus

areas respectively. Note that the accretion rate on the vertical axis spans 30 orders of magnitude. Note especially that the lowest ($\eta \ll 1$) radiative efficiency flows have both higher and lower accretion rates than the thin disks.[221] Higher accretion occurs when the gas density becomes so high that emitted radiation is trapped and cannot escape before falling into the black hole. The radiation pressure dominates over the gas pressure. The accretion rate may now surpass the Eddington critical rate, although the luminosity will remain sub-Eddington. Such structures are found among large-mass, luminous, high-z AGNs. Lower accretion occurs when the gas has low density and is unable to cool. Almost all the energy resides in the ions, and the coupling to electrons is so weak that very little radiation is emitted. Low accretion disks far outnumber other types in the local universe.

The literature is inconsistent with the features ascribed to radiatively inefficient disks. The low accretors are believed to be geometrically thin by some, geometrically thick by others. Some use the terms ADAF (Advection Dominated Accretion Flow) or RIAF (Radiatively Inefficient Accretion Flow) to describe all disks with $\eta \ll 1$, while others restrict the terms ADAF and RIAF to mean only those flows with low accretion. The literature is also inconsistent with the usage of the term *thick disk.* Some use the term in similar fashion to Narayan and Quataert. Others distinguish between a *thick* disk ($M' \gg M'_{\mathrm{Edd}}$) and a *slim* disk ($M' \approx M'_{\mathrm{Edd}}$), where the slim disk is no longer considered radiatively inefficient. The myriad inconsistencies produce a bewildering labyrinth that easily leads the reader astray.

The two variables associated with accretion disk geometry that are most relevant to EGADS are the accretion rate and the radiative efficiency. The following supplemental definitions are offered for better focus on these variables.

α-Mode: very high rate of accretion with low radiative efficiency ($\eta \ll 1$)
Slim-Mode: high rate of accretion with normal radiative efficiency ($\eta \sim 0.1\text{-}0.3$)
β-Mode: very low rate of accretion with low radiative efficiency ($\eta \ll 1$)

α-Mode accretion theory developed from the idea that advection may become relevant when H/R is no longer negligible. The rate of energy dissipation is not related to the angular momentum loss in a simple manner as it is in a thin disk. The viscosity is a more complicated calculation due to the second dimension in the sheer stress tensor. Begelman and Meier (1982) calculated that, "for supercritical accretion ($M' \gg M'_{\mathrm{Edd}}$), most of the radiation is dragged into the hole while the observed luminosity saturates at a fraction of L_{Edd}." [222] Collin *et al.* (2002) found it highly probable that either a large fraction of quasars must accrete at Eddington or super-Eddington rates through geometrically thick disks or that the optical luminosity comes from something other than the disk. [223] The large accretion rates associated with thick disks make α-Mode the likely geometry for EGADS.

Slim-Mode accretion represents a regime between the standard thin disk and α-Mode thick disk. Abramowicz *et al.* (1988) offered a self-consistent treatment of accretion disks assuming optically thick cooling with (thermal bremsstrahlung) electron scattering and free-free opacities. Thermally unbalanced elements expand vertically across the disk thickness more slowly than being advected away from their radial location by the inward drift. The solutions yield thin disks for small accretion rates and slim disks for moderately super-Eddington accretion rates. [224] Luminosity increases only logarithmically with the accretion rate. [225] According to some slim-disk models, the total luminosity is given by

$L = c_1\left(1 + c_2 \ln M'\right) M_{BH}$, where c_1 and c_2 are constants. Slim disks can extend to the black hole horizon, while thin disks terminate at the ISCO. This fully consistent theory of optically thick disks with significant advective cooling establishes another paradigm in accretion disk geometry. Slim-mode disks stay radiatively efficient as the disk bulges to vertical heights of $H/R \sim 0.3$.

β-Mode accretion received little attention in the literature until Ramesh Narayan advanced it in a series of papers in the 1990's.[226] [227] [228] [229] Other models soon followed as the relevance of his work in ADAFs became apparent. Most notable of the derivatives were Advection Dominated Inflow-Outflow Solution (ADIOS) and Convection Dominated Accretion Flow (CDAF). The ADIOS solution has most of the inflow reversed at small radii and driven away in the form of a wind.[230] CDAF is a steady state structure in which the inward transport of angular momentum is exactly balanced by the outward transport.[231] ADAF is radiatively inefficient because most of the energy goes into the black hole. ADIOS and CDAF are radiatively inefficient because the non-accreted mass does not convert its energy into radiation. None of these models produce the necessary torrent of inflow predicted by the EGADS environment. It is relevant, however, that estimates of β-Mode mass accretion rates could be seriously in error.

The best approximation to the nature of AGNs must come from matching observations to theory. Figure 5.5 provides a significant picture of the observed quasar space. Figure 5.10 provides a rough composite of accretion disk theory. The relative precision of the former must be assessed against the latter. Yue Shen exposed the dangers of drawing conclusions from data limited by selection effects. The danger is exacerbated by β-mode disks with low accretion and α-mode disks whose luminosity is rendered weak by radiative inefficiency and heavy cloud cover. Since the accretion rate is related to the observed luminosity *via* η, the accuracy of L/L_{Edd} must come into question whenever efficiency is not near $\eta = 0.1$, the value commonly used in thin accretion disk theory. There exists large variance in η for slim-mode, α-mode and β-mode disks.

Netzer and Trakhtenbrot (2014) (hereinafter NT14) used a large sample of SDSS AGNs to show how adjusted values of η will alter the standard estimations of bolometric luminosity, L_{bol}[standard].[232] Figure 5.11

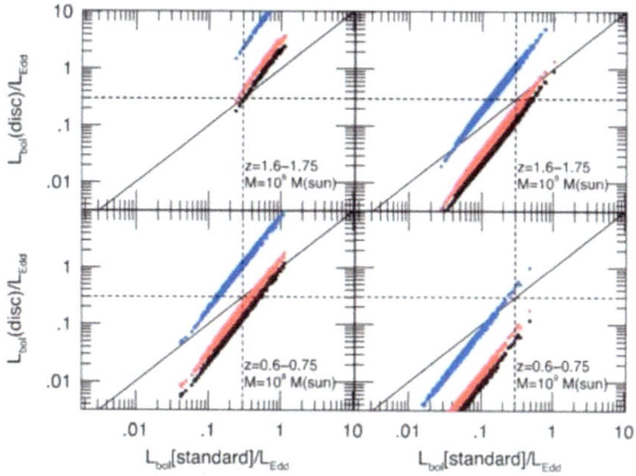

Figure 5.11 – L_{bol} adjusted for three alternate values of η © 2014 Oxford University Press

illustrates the deviations that occur for four combinations of redshift and black hole mass ($z = 0.6$-0.75, $z = 1.6$–1.75, $M_{BH} = 10^8$, and $M_{BH} = 10^9$). The black, red and blue objects have radiative efficiencies of $\eta = 0.038$, 0.057 and 0.321 (spins $\alpha = -1$, 0 and 0.998), respectively.* L_{bol} [disc] represents the corrected bolometric luminosity. If an object lies above a diagonal line, its L_{bol} has been underestimated by standard thin disk theory. Objects in the upper right quadrants (delineated by hash lines) of each z-M_{BH} combination represent the slim disk population.† Several conclusions were formed: (i) The distribution of L_{bol} in the TN12 population is broader and more uncertain than initially assumed; (ii) many type-1 AGNs must contain fast-rotating black holes; (iii) the fraction of slim AGN disks increases with redshift (Figure 5.11 shows that more than 90% of the $z = 1.65$-1.75, $M_{BH} \sim 10^8$ objects are slim-mode). In a survey of luminous unobscured AGNs at $1.5 < z < 3.5$, Trakhtenbrot (2014) showed that it is the most massive BHs that have high radiative efficiencies and spins (see Figure 5.12).[233] This suggests that smaller M_{BH} quasars are dimmer and less likely to meet the flux limit of a given survey. If so, high-z, α-mode AGNs will be well hidden.

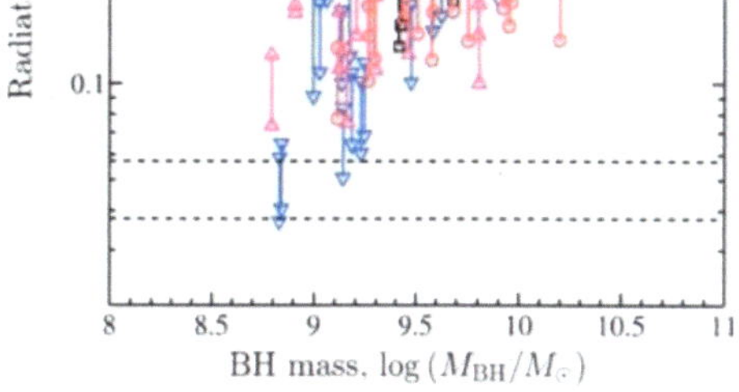

Figure 5.12 – η vs. M_{BH} at high-z

King (2003) predicted that BHs gain most of their mass in phases of super-Eddington inflow.

* See Table 4.3 for conversion of spin to radiative efficiency.

† NT14 arbitrarily defines a slim disk by $M'/M'_{Edd} > 0.3$.

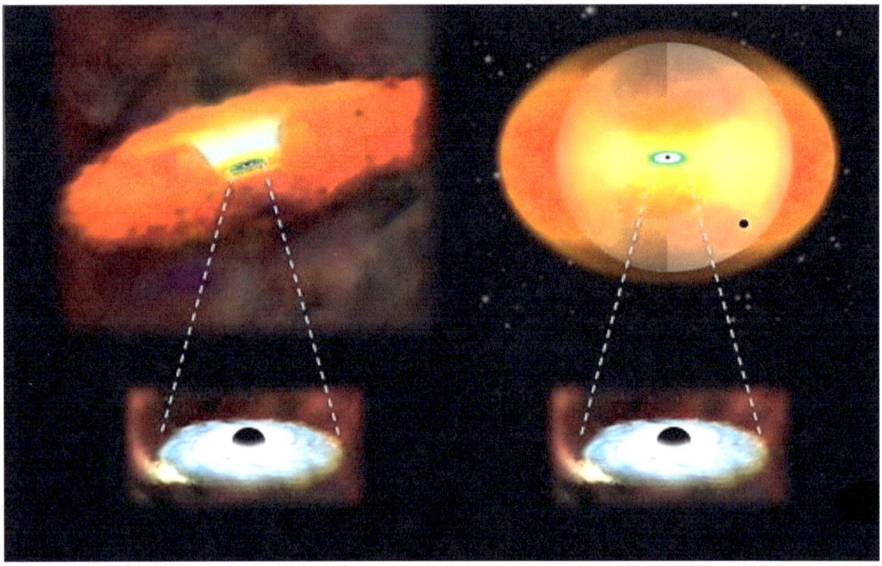

Figure 5.13 – AGN obscuration by torus (left) and galactic merging (right)

Since relatively few active galactic nuclei are observed in such phases, they must be either obscured or at high redshift.[234] The classical AGN unification model predicts that the torus should tilt sufficiently to provide an unobstructed view in some cases. The riddle was resolved in 2017 using NASA's NuSTAR telescope to study 52 AGN in different merger stages in the hard X-ray band. The study revealed that the AGNs become *completely* covered by material funneled from the galactic scale to the inner parsecs during late stages of galaxy mergers. Peak obscuration occurs when the two merging galaxies reach a separation of 0.4 to 10.8 kiloparsecs.[235] The study further supports the theory that merger events are what give rise to the loss of angular momentum (required to incur large-scale infall). The difference between the torus geometry and recent observations is illustrated in Figure 5.13. The closed environment of these objects is very different from that predicted by the AGN unification model. Their geometry appears to substantiate EGADS prediction #4.

α-mode, β-mode and slim-mode disks require different methods of analysis from those used for thin disks. Adjustments are necessary to calculate their temperature, luminosity and spectrum. Previous calculations have come under increasing suspicion. Consider, for example, how NT14 shows the inaccuracies of the outmoded L_{bol} approximations. The AGN unification model has failed to predict the geometry of α-mode super-accretion, while EGADS does accommodate it nicely. It is apparent that an updated model is necessary for an understanding of *all* active galactic nuclei.

Magnetohydrodynamics

It has been a quarter of a century since Igor Novikov spoke to a gathering of the world's most distinguished astrophysicists on the subject of black holes. In the discussion group that followed, Margaret Burbidge posed a highly relevant question.

> (M. Burbidge): Can you imagine the importance of considering magnetic fields and magnetohydrodynamics. This is usually ignored in these discussions.
>
> (Novikov): Yes, relativistic magnetohydrodynamics is very important for discussion of the processes in the neighbourhood of the black hole. Theoreticians, though, start their discussions with simple things, for simplicity. Only recently were the electrodynamics and magnetodynamics of black holes developed.[236]

Novikov presaged the shortcomings of the early thin disk models that did not consider the effects of the magnetic field. Until that time estimations of AGN properties were satisfied on the basis of characteristic support from observations in the local universe. Once the necessity for alternate AGN disk theories became apparent, magnetohydrodynamics (MHD) became a serious issue. The initial focus of MHD assumed force-free magnetic fields limited by time-steady, azimuthally symmetric configurations. Krolik (1999) exposed the limitations of such simplifying assumptions. Unless the plasma somehow suffers tremendous loss of energy through radiation, the randomly accreting matter must transfer a significant fraction of its kinetic energy to the magnetic field.[237] The changing field strengths will have an impact on the flow of energy and angular momentum. Further knowledge of AGN accretion dynamics thus requires measurement of the magnetic field. There is no way to measure the magnetism directly, but it is possible in some cases to make an estimation by means of simulation.

The X-ray emission lines in AGN spectra are assumed to arise in the hot inner regions of the accretion disks in the vicinity of the black holes. The iron K_α emission line (6.4 keV) typically demonstrates a two-peak profile with a high blue peak, a low red peak, and an elongated red wing descending to ground level. Any strong magnetic field in this region will

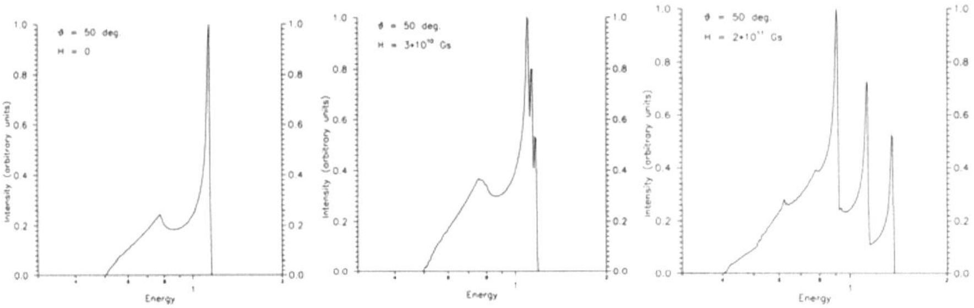

Figure 5.14 – Simulated K_α emission line at three different magnetic field strengths

split these lines into several distinguishable components. Zakharov *et al.* (2003) developed a procedure to estimate an upper limit of the magnetic field in the assumptive region of K_α emission. Simulation of the formation of typical line profiles is performed with different values of B_{BH}, the magnetic field strength near the black hole. When the distortion in the simulated profile matches the observed profile, the value of B_{BH} is established.[238] Figure 5.14 shows the profiles of the K_α spectral line (all at a disk inclination angle of 50°) for three different values of B_{BH}.* The left profile illustrates a null magnetic field. The center profile shows the minimum magnetic field value at which it is possible to observe the line splitting. The right profile has a seven-fold magnetic field increase over the center profile. Notice the clear distinction in linewidth between the latter two profiles.† The spectrum of Seyfert galaxy MCG 6-30-15 ($z = 0.00758$) is matched by a simulated profile with $B_{BH} \sim 10^5$ G (gauss).

The mechanism for the generation of the large magnetic field remains unsolved. A popular theory holds that the kinetic energy of the accreting gas transforms into magnetic energy while in the process of heating the accretion disk. The transformation is known as magnetic coupling (MC). Zhang, Lu and Zhang (2005)[239] probed the magnetic field of 143 low-z AGNs on the basis of the MC model. A negative correlation was found (see Figure 5.15) between black hole mass and magnetic field strength given by the equation

$$\log B_{BH} = 9.26 - 0.81 \log M_{BH} \ .$$

---

* $B_{BH} = \mu_0 H$, where $\mu_0 = 4\pi \times 10^{-7}$ weber/ampere-meter.

† The simulated linewidth depends on the accretion disk model. Since the analysis is done on low-z AGNs, the thin disk model provides reasonable accuracy.

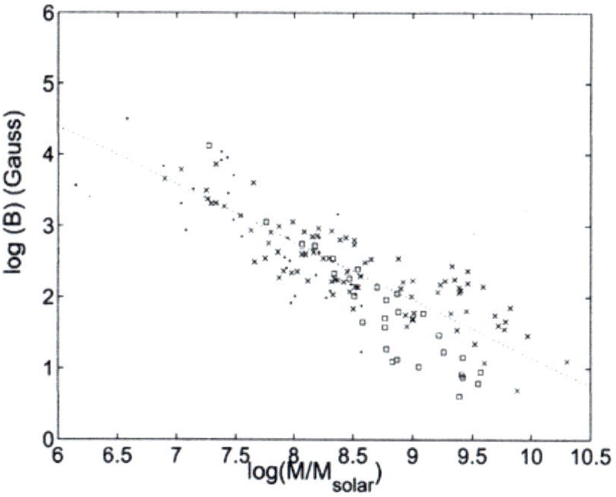

Figure 5.15 – Correlation of B_{BH} versus M_{BH}

More recent studies provide further clues to the nature of the AGN magnetic field and its relative strength. Silant'ev *et al.* (2013) showed the positive correlation of B_{BH} to the accretion rate.[240] Mineshige *et al.* (2002) predicted much larger magnetic influence on disks with low radiation efficiency.[241] Lai (2015) signaled the great importance of the merger remnant wherein the magnetic field can be amplified by differential rotation.[242] Everything suggests that EGADS induces stronger magnetic fields than those found in the low-*z* objects measured by Zakharov or Zhang. The Gedankenexperiment evidently includes the following corollary.

EGADS Corollary

The dark halo merger induces a significant increase in the magnetic field strength of all associated active galaxies.

It is not necessary to look beyond our own Galaxy to find objects with extreme magnetic fields. Neutron stars attain $B \sim 10^{11}$-10^{13} G; magnetars, $B \sim 10^{14}$-10^{15} G. Highly compressed magnetic lines in these objects (caused by star implosion*) produce the extremely powerful fields over which conductive particle/antiparticle pairs (as well as other exotic particles) attain angular velocities near the speed of light. The atomic structure of matter becomes drastically modified in fields where $B \gg 10^9$ G. The atom becomes asymmetric, and the binding energy and oscillator strength become dependent on velocity.[243] The Coulomb force also binds the electrons more effectively along the magnetic field direction.[244] Skjervold and Ostgaard (1984) used the Thomas-Fermi model to approximate ground-state

* A normal star with initial magnetic field of 10-100 G will attain $B \sim 10^{12}$ G when it collapses into a neutron star.

Ground-state binding energies –E [keV] for carbon, oxygen, silicon and iron atoms in superstrong magnetic fields according to the Thomas-Fermi model

B [10^{12} G]	0.1	0.5	1	2	3	4	5	10	50
–E(C)	1.54	2.93	3.86	5.09	5.99	6.72	7.35	9.70	18.46
–E(O)	2.58	4.91	6.48	8.55	10.06	11.28	12.34	16.28	30.99
–E(Si)	7.06	13.45	17.74	23.41	27.54	30.89	33.78	44.57	84.85
–E(Fe)	21.53	40.98	54.07	71.35	83.91	94.14	102.93	135.82	258.56

Table 5.2 – Ground-state binding (keV) energies in strong magnetic fields

binding energies and ionization energies for various elements in field strengths of $10^{11} < B < 5 \times 10^{13}$ G.[245]

Table 5.2 provides the calculated binding energies for carbon, oxygen, silicon and iron. The energies at $B \sim 10^{12}$ G are found in column 4. Not shown is hydrogen with the calculated binding energy of 0.15 keV (as opposed to 0.0136 keV in a null B field). Note the near linear growth of the binding energy as a function of B. This happens because electrons move closer to the binding nucleus as the magnetic field strength increases. Note also that the binding energy increases with increasing atomic number Z. Table 5.3 shows that the calculated ionization energies also increase substantially. Linear extrapolation of the data reveals that the approximate minimum field strength required to observe the HIL blueshift is greater than $B = 10^5$, but it is also well below $B = 10^9$ where the modified atomic structure deforms normal spectral waveforms.

The Alfven radius, R_A, is another important factor in strong magnetic fields. R_A is the radial distance at which the magnetic energy density E_{mag} is equal to the kinetic energy density E_{kin} of the infalling gas. Accreting matter falls radially inward as long as $E_{mag} < E_{kin}$. As the accretion approaches R_A, it becomes increasingly deflected under the influence of the

Ionization energies I [eV] for carbon, oxygen, silicon and iron atoms in superstrong magnetic fields according to the Thomas-Fermi(-Dirac) model

B [10^{12} G]	0.1	0.5	1	2	3	4	5	10	50
I_Z(C)	18	33	44	58	69	77	84	111	211
I_Z(O)	16	31	40	53	63	70	77	101	193
I_Z(Si)	14	26	34	45	53	60	65	86	164
I_Z(Fe)	12	22	29	38	45	50	55	73	138
I_{Z-1}(C)	61	117	154	204	239	269	294	388	738
I_{Z-1}(O)	55	104	137	181	212	238	261	344	655
I_{Z-1}(Si)	45	85	112	148	174	195	213	282	536
I_{Z-1}(Fe)	37	70	93	122	144	161	176	233	443

Table 5.3 – Ionization (eV) energies in strong magnetic fields

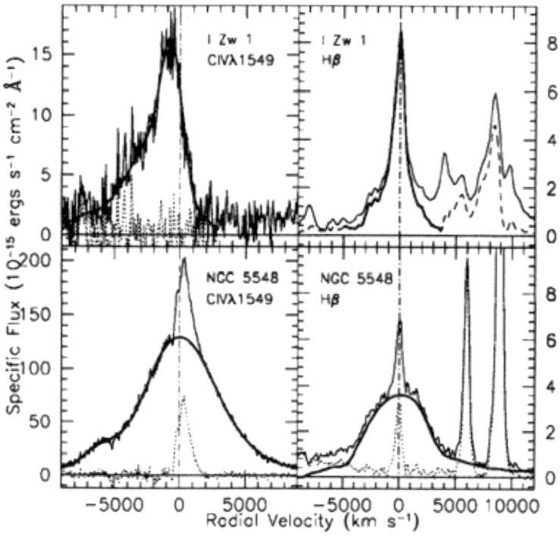

Figure 5.16 – C_{IV} and $H\beta$ line profiles for sources I Zw 1 and NGC 5548

magnetic field. Little such deflection occurs for low-*z* AGNs where R_A exists near the black hole horizon. For AGNs with sufficiently large E_{mag}, deflection will begin further out in the accretion disk. Since the temperature of the disk has radial dependence of $T \sim R^{-3/4}$, the higher ionization occurs at smaller radius where the larger angle of deflection reduces the inward velocity. Some combination of diminished inward velocity and higher ionization energies will cause an observed blueshift of HIL (high ionization lines) over LIL (low ionization lines) in the spectrum.*

Observations show HIL are systematically blueshifted with respect to LIL in many AGNs. Some dependence on luminosity and/or redshift seems apparent, but it has not been possible to untangle the relationships. For bright, high-*z* quasars the blueshift is ~1000 km s^{-1} or greater.[246] Tytler and Fan (1992) found that blueshift increases roughly with energy level.[247] Sulentic *et al.* (2000c) confirmed that essentially all radio quiet (RQ) sources show some $C_{IV}\lambda1549$ blueshift relative to $H\beta\lambda4861$. Figure 5.16 presents a comparison of the C_{IV} and $H\beta$ line profiles for objects I Zw 1 ($z = 0.061$) and NGC 5548 ($z = 0.016$).[248] The dark lines trace the broad-line component (BC), and the dotted lines show the residual after BC subtraction. Note the clear differences in the Balmer profiles and the kinematic decoupling of the C_{IV} and $H\beta$ profiles in object I Zw 1.

Explanations for HIL blueshifting fall into two broad categories. One possibility is that HILs arise in winds flowing *outward* above the accretion disk. Most early AGN models make this assumption. The other possibility is that *inward* velocity decreases by some means of scattering. The best method to determine the direction of flow is through velocity-resolved reverberation mapping (VRRM). The figure on the right shows the VRRM

* HILs with shorter wavelengths will also undergo *less redshift* than LILs due to gravitational potential. See Chapter 4, **Proximity and Mass of the Black Hole.**

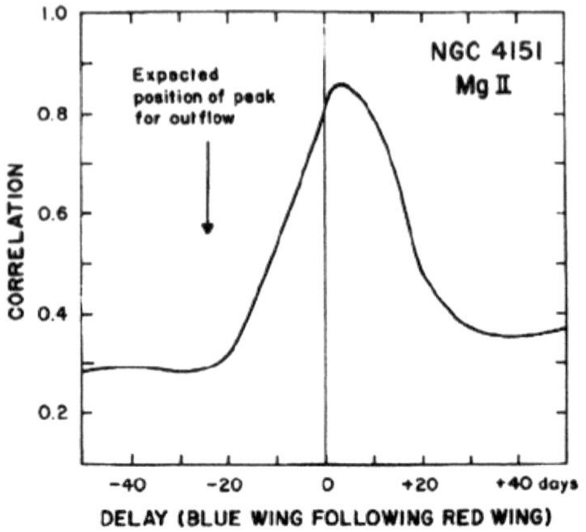

Figure 5.17 – NGC 4151 Mg_{II} variance in time
© 2009 Elsevier

cross-correlation function for the blue and red wings of the (NGC 4151) Mg_{II} line as a function of time delay.[249] For outward flow the blue wing will respond to continuum changes first. This is not what is seen in Figure 5.17. The lag of the blue wing behind the red wing indicates an inflow of the gas dominated by gravitational and/or magnetic influence. This result has been found repeatedly for many lines in many objects. Further statistics of line widths also imply that there must be a substantial component of velocity *perpendicular* to the orbital plane. Gaskell and Goosmann (2016) supplemented the VRRM analysis with other evidence to harden the case for inflow.[250] If blueshifting is not always the result of gas outflow, a consistent model that also includes inflow becomes necessary.* Scattering off an inflowing medium is a possibility, but it does not satisfactorily explain the anticorrelation of Hβ redshifting with C_{IV} blueshifting. The easier explanation comes from AGNs with magnetic field strengths greater than $B = 10^5$ G, the upper limit employed by most current models. Observational evidence is consistent with stronger B and EGADS theoretical prediction #5.

Reionization of the Intergalactic Medium

Much of the initial research in QSO spectra endeavored to determine whether Lyα absorption lines are caused by intervening clouds in the intergalactic medium (IGM).† Popular theory now distinguishes between two types of absorbers: low column density absorption by neutral hydrogen distributed in the web of dark matter; high column density absorption by intervening galaxies and proto-galaxies. These *external* absorbers over a wide range of redshifts establish the best argument for the cosmological hypothesis (CH). Once AGNs are well established at their cosmological

* The dominant motions of most of the BLR are orbital and turbulent.
† See Chapter 2, **Lyman-Alpha Absorption**.

distances, research can turn around and study the IGM in further detail. One fundamental cosmological question is when the primordial elements reverted from a neutral state back into ionized plasma. The process that caused neutral matter in the universe to reionize is known as *reionization*. The period when this occurred is known as the *reionization epoch*.

One method to study reionization is to measure the neutral hydrogen fraction, $N_{HI} = n(H_I)/n(H)$, in the spectra of distant quasars. The best test for neutral gas as a component of the IGM uses the mapping between distance and observed frequency in order to identify Lyα absorption caused by diffuse matter spread over a wide redshift range.* This is known as the *Gunn-Peterson test.* A tiny neutral density fraction of $N_{HI} = 10^{-4}$ is sufficient to saturate the absorption (*i.e.*, spectra become insensitive to density fractions greater than 10^{-4}). In the expanding Universe characterized by a uniform IGM containing neutral hydrogen, the continuously redshifted Lyα resonance will lead to a Lyα absorption *trough* blueward of the source frame Lyα line. [251] The reionization epoch may be assumed to approach its completion at that redshift where the Lyα trough transitions to the Lyα forest. Based upon spectroscopy of quasar SDSS 1044-0125 ($z = 5.73$), Djorgovski *et al.* (2001) suggested the trailing end of hydrogen reionization exists somewhere near $5.2 < z < 5.8$. Figure 5.18 shows the dramatic increase found in the density of the Lyα forest at λ7550. [252] Results based upon spectra of QSOs at $z > 6$ also show that the effective optical depth increases

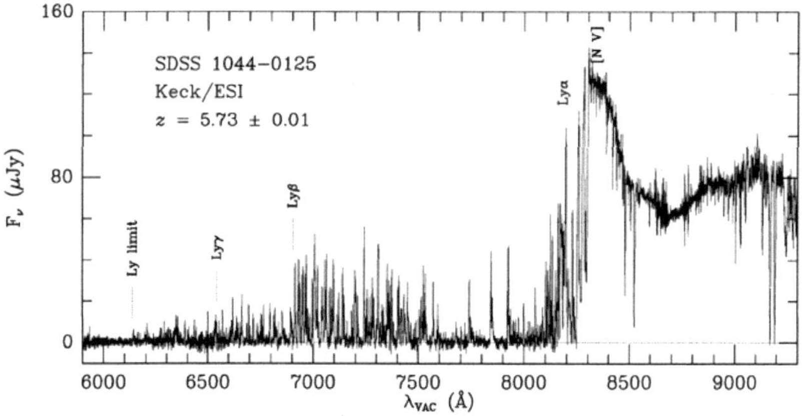

Figure 5.18 – Lyα forest (SDSS 1044–1025)

* Such absorption is expected to have optical depth $\tau = 1.7 \times 10^{11} N_{HI}(z)/(1 + z)^{3/2}$.

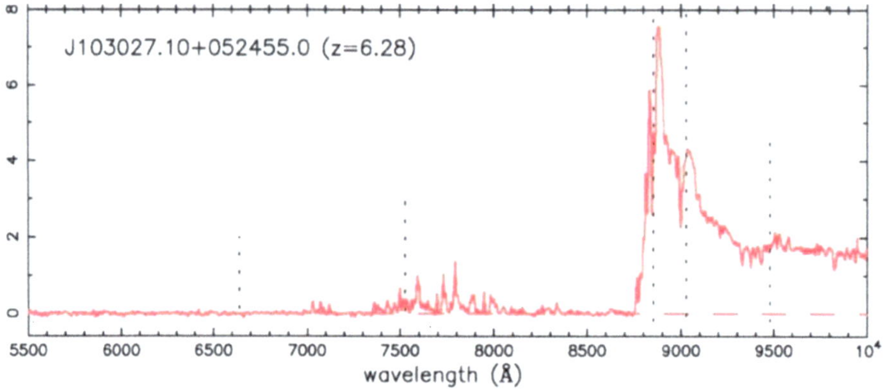

Figure 5.19 – Lyα trough (SDSS J1030+0524) (Black dashed lines indicate the locations of expected prominent emissions)

rapidly with redshift. Becker *et al.* (2001) were the first to report a complete Gunn-Peterson trough caused by neutral hydrogen in the intergalactic medium. Figure 5.19 reveals the spectrum of quasar SDSS J103027.10+052455.0 ($z = 6.28$) which contains a flux level consistent with zero in a ~300 Å region immediately blueward of the Lyα emission line. The authors are careful to state: "Only a small fraction of neutral hydrogen component in the IGM is needed to have $\tau_{GP} \gg 1$ and give rise to a complete Gunn-Peterson trough. Therefore, the existence of a Gunn-Peterson trough by itself does not guarantee that the object is observed prior to the reionization epoch." [253] Any notion that zero-flux troughs would exist in every $z > 6.2$ quasar was dashed by Oh and Furlanetto (2005), who revealed that flux is visible in the Lyα and Lyβ troughs of SDSS J1148+5251 ($z = 6.42$). [254] Fan *et al.* (2006) then published their landmark study of 19 quasars at $5.74 < z_e < 6.42$ that further confirms the significant cosmic variance along different lines of sight. [255]

Despite all of the progress in recent years, the reionization epoch remains shrouded in mystery. First, it remains unclear how quickly the IGM becomes neutral above $z = 6$. Second, many Lyα forest systems are found enriched in heavy elements. Third, the energy source for the ionization remains a puzzle. The source must exist (and remain) in sufficient magnitude lest IGM protons and electrons recombine. Dwarf galaxies and quasars do not exist in adequate numbers to reionize the IGM. Population III stars are now considered the most likely source, but they have never been directly observed outside our Galaxy. Fourth, observations of the CMB stand in conflict with measurements of the neutral hydrogen fraction. *Planck 2013 Results* documents an instantaneous reionization at redshift $10.2 < z_{reion} < 12.4$.[256]

The Gunn-Peterson test can be based on absorption by atoms other than hydrogen. Neutral helium (He_I) absorption can cause a flux decrement at 584 Å < λ < 584 Å $(1 + z_e)$; singly ionized helium (He_{II}) absorption, at 304 Å < λ < 304 Å $(1 + z_e)$. The reionizations of primordial H_I and He_I are very similar. Both will ionize from a photon with energy greater than 24.6 eV, and both have comparable recombination rates. However, He_{II} ionizes at the higher energy of 54.4 eV and recombines more rapidly. He_{II} reionization ($He_{II} \rightarrow He_{III}$) thus takes place at a later cosmic period, and it can be decoupled from the other two species to provide further insight into the reionization epoch.[257] Analogous to hydrogen, the optical depth of He_{II} is used to gain estimations of the reionization. Although in greater abundance than neutral hydrogen, ionized helium imposes many more challenges in measurement: the short wavelength of He_{II} Lyα absorption can only be observed at $z > 2.5$ even with the far UV bands accessible to Hubble Space Telescope; QSOs at $z_e < 3.1$ are not clear of geocoronal (H_I Lyα and O_I λ1302) contamination; most QSOs at $z_e > 3$ are completely unobservable due to strong photoelectric absorption in the numerous Lyα systems intercepted along the line of sight. In their report of the first detected He_{II} Lyα Gunn-Peterson trough,* Jakobsen *et al.* (1994) revealed that a typical line of sight to $z \sim 3$ intercepts an average of ~400 H_I Lyman forest clouds and ~6 H_I Lyman limit clouds.[258]

Until recently only a very small number of suitable quasar sightlines were available for helium reionization study. A more efficient method to locate appropriate sightlines became available with the advent of the Galaxy Evolution Explorer (GALEX) all-sky UV survey. Syphers *et al.* (2009) cross-matched the bright GALEX sources with existing SDSS $z > 2.8$ sources and discovered twenty clean sightlines.[259] Further improvement came with the installation of the Cosmic Origins Spectrograph (COS) aboard *Hubble Space Telescope* in 2009. The increased observation power allows high resolution of IGM helium down to $z = 2.5$ with flux from lower redshift quasars. Syphers *et al.* (2012) were then able to report an addi-

---

* Interestingly, the reported He_{II}Lyα trough in Q0302-003 ($z = 3.286$) precedes the first reports of H_ILyα troughs.

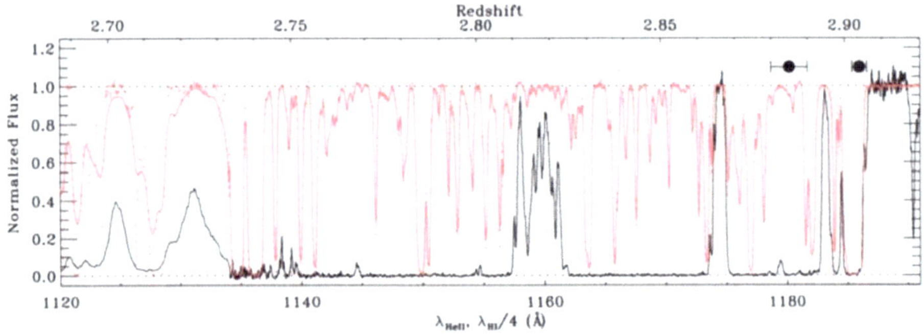

Figure 5.20 – Normalized COS He_{II} (**black**) and VLT H_I (**red**) flux of quasar HE 2347-4342 (Solid black bullets denote the original z_{QSO} ~ 2.885 and revised z_{QSO} ~ 2.904 systemic redshifts)

tional 13 He_{II} sightlines at $2.82 < z < 3.77$ from the COS observations.[260] Shull *et al.* (2010) compared the *HST*/COS observations of quasar HE 2347-4342 with older spectra from VLT (Very Large Telescope) and *FUSE* (*Far Ultraviolet Spectroscopic Explorer*).[261] Figure 5.20 shows an overlay of the COS He_{II} Lyα absorption (black) with the VLT H_I Lyα absorption (red). Red H_I wavelengths are divided by 4 for easier comparison with the black He_{II}. The $\eta = He_{II}/H_I$ optical depth ratio varies widely between 5 and 500.* Figure 5.21 compares the COS/G130M medium resolution (black) and COS/G140L low resolution (blue) observations with data from *FUSE* (green). Three long troughs (4-10 Å or 25-60 Mpc comoving distance) of strong He_{II} absorption (optical depths $\tau_{HeII} > 5$) are seen clearly for the first time between $z = 2.75$ and 2.90. The narrow windows of partial flux transmission interspersed with the broad He_{II} absorption troughs indicate ongoing He_{II} reionization. The more sophisticated COS observation shows that the epoch of He_{II} reionization evolves more *slowly* and ends at $z < 2.7$, in contradiction to earlier estimates of $z_{reion} > 3$.

The fluctuating He_{II} must occur from variations in IGM density or ionization levels. The underlying density field is traced by the coeval H_I Lyα forest (see Figure 5.20). When $\eta = He_{II}/H_I$ is constant, the increase in $\tau_{HeII}(z)$ will be due to density evolution in the IGM. For an interval of constant H_I, the fluctuation in He_{II} is caused by variation in the ionization level. The HE 2347-4342 optical depth ratio of $\eta = 5$-500 seems extreme in

* $\eta = 5$ implies an extreme hard quasar UV background; $\eta = 200$ may be a soft galaxy-dominated UV background.

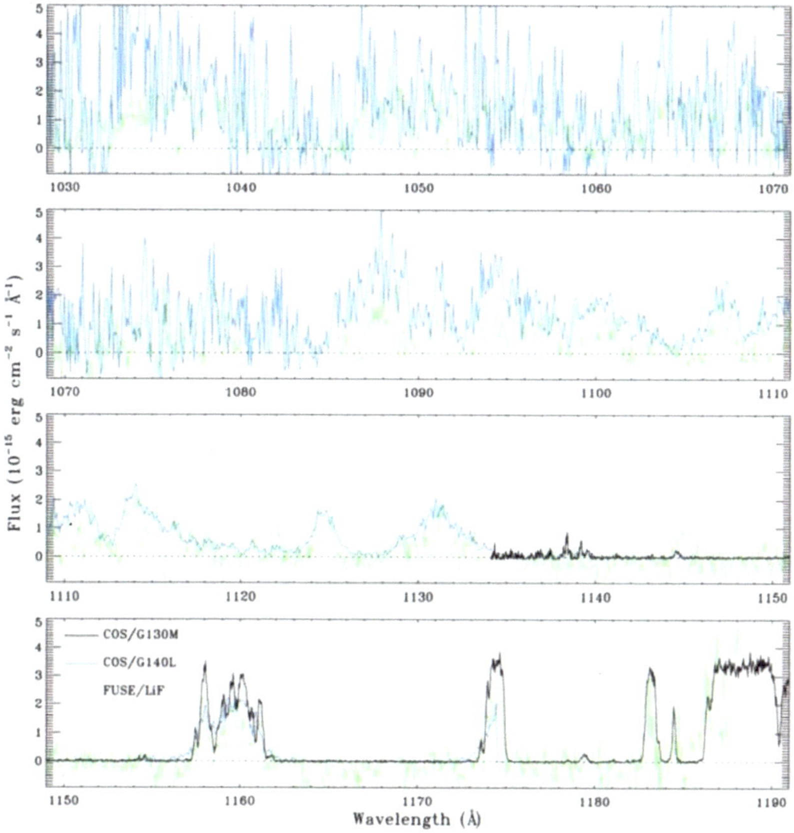

Figure 5.21 – Comparison of flux between COS (blue & **black**) and *FUSE* (green) of quasar HE 2347-4342

any case. Analysis of more GALEX UV sightlines would help clarify the incongruity.

A group of astrophysicists recently established the Helium Reionization Survey (HERS), a campaign with the objective of addressing all technical aspects of He_{II} reionization. Their vanguard study (Worseck *et al.*, 2016) of 17 UV-transmitting quasars ($2.3 < z < 3.5$) reveals distinct challenges to existing scenarios of when and how the IGM became ionized.[262] A strong sightline-to-sightline variance exists in the effective optical depth of He_{II} Lyα absorption, $\tau_{eff,\,HeII}$. Reionization is ongoing at $z < 2.7$ and was well underway at $z = 4$. This is in striking conflict with predictions of He_{II} reionization driven by luminous quasars. Furthermore, He_{II} absorption is not well correlated with the coeval H_I Lyα forest. The significant dispersion of He_{II}/H_I reveals that the fluctuating He_{II} absorption patterns are primarily due to large variations in ionization level.[263] This now presents the challenge of finding the source of hard photons that ionize He_{II} varyingly between $z < 2.7$ and the early ($z > 4$) universe.*

Recent observations of H_I and He_{II} Lyα absorption challenge the existing model of reionization. Quasar sightlines reveal an extended and 'patchy' process. Although this does not compromise the consistency of the cosmological hypothesis, new difficulties have arisen. The easy solution is afforded by EGADS theoretical prediction #6.

Summary

Six theoretical predictions are offered in support of EGADS: 1) luminous bridges will be seen between objects of disparate redshift; 2) black hole mass will not evolve according to the standard model; 3) average BH accretion rate will show a monotonic rise with increasing redshift; 4) AGNs will darken significantly as they become enshrouded during galaxy mergers; 5) high ionization lines will appear blueshifted relative to low ionization lines; 6) strong variances in the Lyα forests will be seen in the sightlines to distant quasars. Recent observations have supported all six predictions.

* New models now must overcome many caveats in order to explain the necessary ionizing photons. Hao *et al*. (2015) offered (hypothetical) mini-quasars as a possibility.[264]

Chapter 6

Distance-Independent Data Analysis

The mystery of AGNs cannot be unlocked until the relatively small dimensions of their active nuclei can be resolved. Existing world-class telescopes lack the capability to do so. The situation will change with the new class of megascopes under construction. The Giant Magellan Telescope (GMT), the European Extremely Large Telescope (EELT), and the Thirty Meter Telescope (TMT) are all scheduled to come online in the 2020s. In addition to these land-based behemoths, the James Webb Space Telescope (JWST) will soon launch into space and begin observations at a Lagrange point one million miles from Earth. The astronomical community waits in eager anticipation to see which old issues will be solved and what new mysteries will arise. A very important issue to resolve is the imprecision of distance measurements. It is not certain to what degree this will be accomplished.

The theory of AGN evolution rests in part on the locations we ascribe to the distant quasars. Various small-scale inaccuracies in distance have been noted for low-z objects. It is possible that yet more severe inaccuracies exist for the entire population, and it is not certain whether megascopes will unravel the matter. A better approach may be to consider AGN evolution with a picture that does not rely on any redshift dependent variables (such as black hole mass M_{BH}, luminosity L, and absolute magnitude M_ν). Spectroscopy provides an abundance of distance independent variables from which to choose. The trick is to find the appropriate variables whose correlations provide a meaningful physical description of the observed object.

Principal Component Analysis

Consider a dataset whose objects are described by ξ defined variables. The number of correlations between any two variables is then given by combinations of ξ taken two at a time, or $\xi!/2(\xi-2)!$. The best means for sorting the correlations into meaningful groups is a statistical tool known as *principal component analysis* (PCA). PCA transforms data into a new coordinate system in which each successive coordinate eigenvector accounts for as much of the remaining variance in the data as possible. All eigenvectors are orthogonal and the number cannot exceed $\xi-1$. The potential for PCA to provide meaningful interpretations of variable emission line properties was illustrated elegantly by Boroson and Green (1992) (hereinafter BG92). Thirteen properties within the spectral range of λλ4300-5700 were considered for eighty-seven low-*z* QSOs.[143] Table 6.1 shows the properties against the first five PCA eigenvectors. The percentage of accumulated variance for the first *n* eigenvectors is indicated at the top of the nth eigenvector column. The derived transform matrix makes it easy to visualize

Property	Eigenvector 1 29.2%	Eigenvector 2 50.9%	Eigenvector 3 61.4%	Eigenvector 4 68.2%	Eigenvector 5 74.2%
M_ν	–0.269	+0.852	–0.087	+0.098	+0.051
Log *R*	+0.566	–0.324	–0.404	–0.148	+0.182
α_x	–0.340	–0.543	–0.242	+0.144	–0.220
EW Hβ	+0.265	+0.153	+0.715	+0.412	+0.314
R λ5007	+0.598	+0.452	–0.510	–0.002	–0.183
R λ4686	–0.033	+0.792	–0.253	+0.186	+0.027
R Fe II	–0.832	–0.078	–0.201	–0.166	–0.145
Peak λ5007	+0.845	+0.269	–0.270	–0.003	–0.125
Hβ FWHM	+0.654	–0.221	+0.235	–0.240	–0.023
Hβ shift	–0.025	–0.229	–0.504	+0.116	+0.773
Hβ shape	–0.173	–0.400	–0.288	+0.732	–0.220
Hβ asymm	–0.679	–0.067	–0.119	–0.233	+0.153
$M_{[O\ III]}$	–0.811	+0.400	+0.007	–0.061	+0.065

Table 6.1 – Eigenvector correlations of low-z QSO emission line properties

M_ν – monochromatic absolute magnitude at λ5500
Log *R* – *log of the ratio of radio to optical flux density (5 GHz / 4400 Å flux)*
α_{ox} – optical to X-ray spectral index
EW Hβ – equivalent width of Hβ including both broad and narrow components
R λ5007 – ratio of EW(O_{III}) to EW(Hβ)
R λ4686 – ratio of EW(He_{II}) to EW(Hβ)
R Fe_{II} – ratio of EW(Fe_{II}) to EW(Hβ)
Peak λ5007 – ratio of the peak height of [O_{III}] λ5007 to Hβ
Hβ FWHM – full width at half maximum of the Hβ broad component
Hβ shift – shift of the centroid at ¾ maximum from the rest wavelength in units of the FWHM
Hβ shape – parameter shape
Hβ asymm – a measure of the shift between the centroids at ¼ and ¾ maximum in units of the FWHM ($\lambda_{c(3/4)} - \lambda_{c(1/4)}$/FWHM). Positive values indicate excess light in the blue wing; negative values indicate excess light in the red wind.
$M_{[OIII]}$ – absolute magnitude of λ5007 defined as $M_\nu = -2.5 \log$ (EW λ5007)

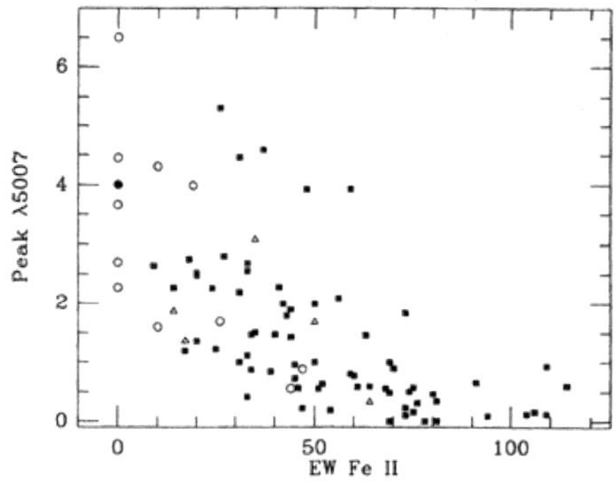

Figure 6.1 – Anticorrelation of $[O_{III}]$ and Fe_{II}

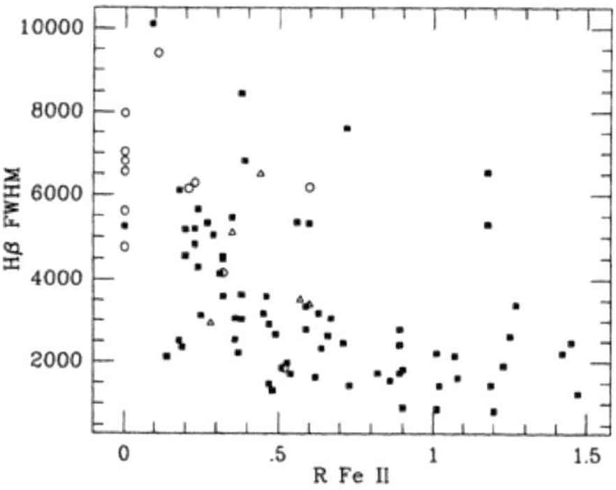

Figure 6.2 – Anticorrelation of Hβ FWHM and Fe_{II}

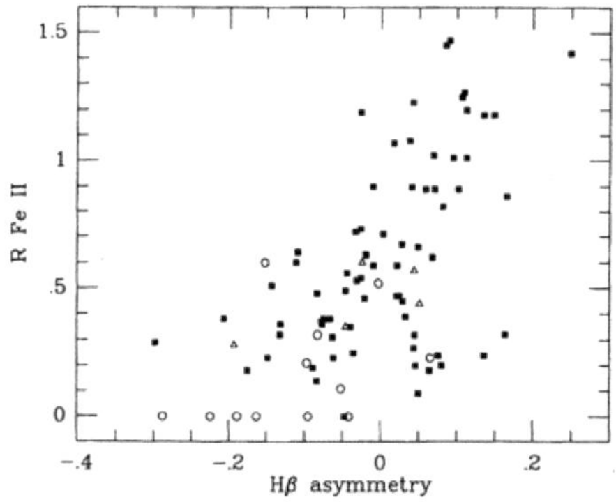

Figure 6.3 – Correlation of Fe_{II} and Hβ asymmetry

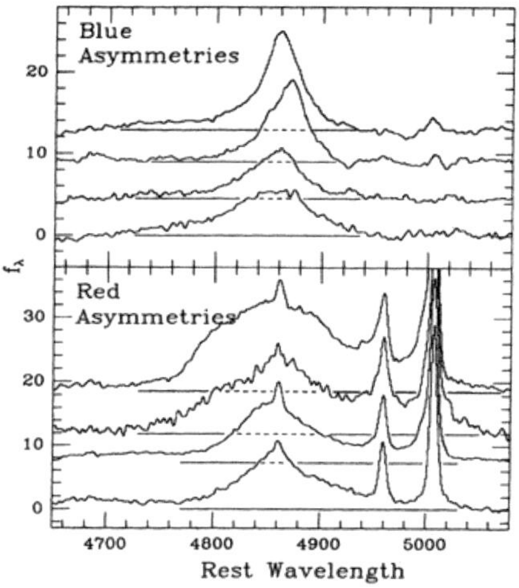

Figure 6.4 – Hβ asymmetry profiles

Key to Figures 6.1 - 6.3: Radio Quiet (RQ) objects are indicated by solid squares; Radio Loud (RL) flat-spectrum objects, by open triangles; Radio Loud (RL) steep-spectrum objects, by open circles.

the most significant correlated properties.

Eigenvector 1 reveals the high (anti-) correlation between R Fe_{II} and Peak λ5007, where R Fe_{II} is tightly correlated to Hβ *asymm*, and Peak λ5007 is tightly correlated to Hβ FWHM. Eigenvector 2 is dominated by the anti-correlation of luminosity (M_ν) and the strength of Fe_{II}. This second eigenvector is removed from consideration here due to the fact that luminosity is a distance-dependent variable. Eigenvectors 3, 4 and 5 are dominated by single properties (EW Hβ, Hβ shape, and Hβ shift, respectively) with little correlation to anything else. Since the latter three account for 23.3%

(74.2%–50.9%) of the variance, their properties are more likely driven by different physical parameters than the first two eigenvectors.

Log R is also an important property to consider because the underlying physical radio process has been studied at great length. The strong (±) correlations in eigenvector 1 of *Log R* to properties Peak λ5007, Hβ *asymm*, and Hβ FWHM provide further constraints on the radio processes of AGNs. Figure 6.1 shows the strong anticorrelation of Peak λ5007 and EW Fe_{II}. The radio-loud (*Log* $R > 1$) and radio-quiet (*Log* $R < 1$) sources are indicated by open circles and solid squares, respectively. It is evident that radio-loud sources do not exist in the presence of strong Fe_{II}. The same phenomenon appears in Figure 6.2 where Hβ FWHM is measured against R Fe_{II}.* Figure 6.3 plots R Fe_{II} against the Hβ asymmetry. The dearth of blueward asymmetry in radio-loud sources provides another important clue. Spectra make this more apparent. Figure 6.4 shows the spectral profiles where either blue asymmetry or red asymmetry exists in Hβ. The contrast is striking. The absence of any noticeable [O_{III}] lines implies a radio-quiet source. Although the sample size was small by current standards, the BG92 study has stood the test of time. All AGNs for which any PCA is conducted reveal strikingly similar property correlations.

Property	Spearman parameter	L_x correlation probability
M_υ	–0.635	>99.9%
Log R	0.403	99.7%
α_{ox}	–0.466	99.9%
EW Hβ	0.229	90.8%
R λ5007	0.036	21.1%
R λ4686	–0.300	97.2%
R Fe_{II}	–0.474	>99.9%
Peak λ5007	0.247	93.1%
Hβ FWHM	0.371	99.4%
Hβ shift	0.052	29.2%
Hβ shape	–0.174	80.0%
Hβ asymm	–0.471	>99.9%
$M_{[OIII]}$	–0.717	>99.9%
EW FeII	–0.304	97.5%

Table 6.2 – Probability of correlation to L_x

The wealth of information acquired from the relatively small band (λλ4300-5700) of observation served to whet the appetite for similar analyses over as much of the electromagnetic spectrum as possible. Michael Corbin took on the task of comparing the BG92 optical emission-line properties with 2 keV X-ray luminosities already recorded in the *Einstein Observatory* database (for objects at $z < 0.5$). His analysis revealed that the X-ray luminosity L_x is another variable correlated with the

* Hβ replaces [O_{III}] as a principal variable in more recent AGN eigenvector analysis for reasons explained later.

BG92 properties. Table 6.2 indicates R Fe_{II}, Hβ asymm, and $M_{[OIII]}$ all correlate with log L_x with a probability > 99.9%.[265] He further found that they do not seem to be influenced by orientation, and they are demonstrated in both radio-loud and radio-quiet objects. Wang, Brinkmann and Bergeron (1996) further revealed that extreme Fe_{II} emission quasars have significantly lower L_x, slightly steeper broadband optical to X-ray spectral index α_{ox}, and a significantly steeper soft X-ray photon index Γ_x. Neither starburst activity nor any simple thermal model can explain the (0.1-2.4) keV X-ray spectra.[266]

Corbin and Boroson next retrieved UV spectra from the archival *HST* and *IUE* observations. Eigenvector 1 of the combined optical/UV data (~1100 to 5500 Å) confirmed that radio-loud (RL) objects are more luminous than radio-quiet (RQ) objects and that the strongest source of variance in the optical lines is the anticorrelation of the $[O_{III}]\lambda5007$ with optical Fe_{II} strengths. The new PCA revealed strong and previously unseen correlations in the C_{IV} profile: C_{IV} asymmetry increases redward with UV luminosity ($L_\nu\lambda1549$) but strongly blueward in higher-z objects; C_{IV} line centroids show blueshifts ~50-200 km s^{-1} (relative to $[O_{III}]\lambda5007$) compared to Hβ redshifts ~75 km s^{-1}; both Fe_{II} and C_{IV} show a strong Baldwin effect, where line strength is anticorrelated to luminosity. Table 6.3 updates the BG92 correlation matrix with C_{IV} properties. Corbin and Boroson (1996) stated, "The correlation of the widths and asymmetries of the UV lines with Hβ argues strongly against models in which they arise in physically distinct regions, outflowing gas for the former and optically thick accretion disk for the latter."[267]

Corbin next focused his attention on the asymmetries and displacements in the line profiles. He showed that the gravitational effect on Hβ

Parameter	log L_ν(1549)	Log L_X/2 keV	EW Lyα	EW C_{IV}	FWHM Lyα	FWHM C_{IV}	Asymm C_{IV}	EW Hβ	EW Fe_{II}	EW λ5007	EW λ4686	FWHM Hβ	Asymm Hβ
log L_ν(1549)	1	0.703	−0.196	−0.263	0.141	0.285	−0.650	0.090	−0.428	−0.127	−0.334	0.259	−0.374
Log L_X/2 keV	0.703	1	−0.155	−0.026	0.105	0.181	−0.336	0.129	−0.480	0.038	−0.182	0.344	−0.393
EW Lyα	−0.196	−0.155	1	0.565	0.124	0.181	−0.110	0.279	−0.352	0.984	−0.169	0.000	−0.003
EW C_{IV}	−0.263	−0.026	0.565	1	0.275	0.118	−0.184	0.437	−0.582	0.588	0.219	0.456	−0.171
FWHM Lyα	0.141	0.105	0.124	0.275	1	0.675	−0.285	0.233	−0.249	−0.240	−0.388	0.466	−0.160
FWHM C_{IV}	0.285	0.181	0.181	0.118	0.675	1	−0.245	0.326	−0.283	−0.318	−0.348	0.486	−0.096
Asymmetry C_{IV}	−0.650	−0.336	−0.110	−0.184	−0.285	−0.245	1	0.028	0.241	0.027	0.151	−0.383	0.422
EW Hβ	0.090	0.129	0.279	0.437	0.233	0.326	0.028	1	−0.069	0.303	0.197	0.196	−0.095
EW Fe_{II}	−0.428	−0.480	−0.352	−0.582	−0.249	−0.283	0.241	−0.069	1	−0.597	−0.068	−0.690	0.532
EW λ5007	−0.127	0.038	0.094	0.588	−0.240	−0.318	0.027	0.303	−0.597	1	0.417	0.199	−0.370
EW λ4686	−0.339	−0.182	−0.169	0.219	−0.388	−0.348	0.151	0.197	−0.068	0.417	1	−0.351	−0.085
FWHM Hβ	0.259	0.344	0.000	0.456	0.466	0.486	−0.383	0.196	−0.690	0.199	−0.351	1	−0.462
Asymmetry Hβ	−0.374	−0.393	−0.003	−0.171	−0.160	−0.096	0.422	−0.095	0.532	−0.370	−0.085	−0.462	1

Table 6.3 – Emission-line and continuum parameter correlation matrix

emissions from the VBLR and ILR could produce the redward asymmetries and redshifts.* However, the blueward shifts of VBLR components require an *additional mechanism in competition with the gravitational redshift.* This fact is made all the more extraordinary in that the peaks of HIL (C_{IV}) lines tend to be blueshifted by ~100 km s^{-1} versus comparable redshifting of the LIL (Hβ) peaks, even though the origin of these lines is the same ensemble of emitting clouds.[268] The origin of the blueshifts remains a matter of debate.†

Radio Loudness and Evolution

The theoretical model of AGN structure is built upon a wide array of observational evidence. PCA is a tool that can further constrain the model and help unravel the mystery of AGN evolution. The *Log R* parameter is most important in this regard. *R* represents the AGN radio loudness defined by the ratio of radio (5 GHz) to optical (4400 Å) monochromatic luminosity. The dividing line between a radio-loud and radio-quiet AGN is customarily set at $R = 10$. A significant dichotomy exists between RL and RQ quasars, the underlying reasons for which remain a matter of debate. Virtually all RL AGNs are hosted by giant elliptical galaxies whose surface brightness profiles show light deficit in the nuclear region, while RQ AGNs reside in less massive, elliptical galaxies whose surface brightness profiles show light excess in the nuclear region.[269] Although most AGNs emit some radio flux, the radio-loud population is only 10% and decreases with redshift. RL AGNs also exhibit an important distinction in terms of their Fanaroff and Riley (FR) morphologies. FR I radio galaxies are edge-darkened sources with low-ionization spectra and weak emission lines. FR II radio galaxies are higher luminosity, edge brightened sources often associated with broad-line emitting quasars. The much fewer FR II sources typically exhibit linear polarization perpendicular to jet direction, suggesting magnetic fields associated with the synchrotron emission.

Research has shown that the RL/RQ dichotomy is due in some part to differences in the central black hole mass, accretion rate, accretion mode, and spin. McLure and Jarvis (2004) found that RL quasars (RLQs) are 45%

---

* See Chapter 4, **Proximity and Mass of the Black Hole**.

† See Chapter 5, **Magnetohydrodynamics**, for discussion of blueshift in the EGADS environment.

more massive than RQ quasars (RQQs) at fixed optical luminosity. Figure 6.5 on the right reveals that the RLQs (indicated by solid black circles) are mostly confined to $M_{BH} > 10^8\ M_{Sun}$. At fixed black hole mass, however, the range in the radio luminosities covers several orders of magnitude. Furthermore, RQQs are indistinguishable from RLQs in terms of orientation and host galaxy properties.[270] It is evident that additional factors must be invoked in order to explain the RL/RQ dichotomy.

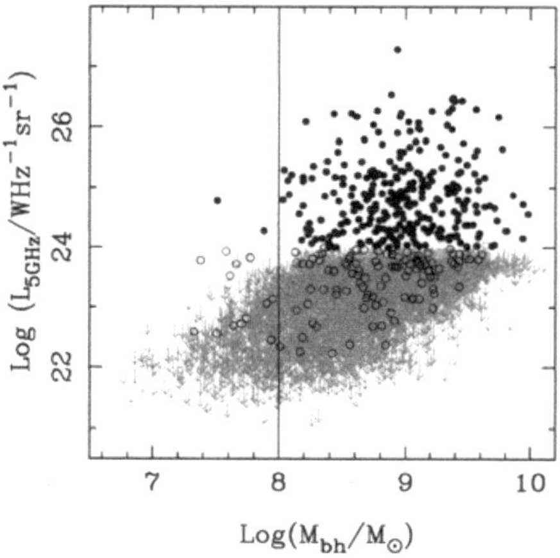

Figure 6.5 – Radio luminosity versus M_{BH} RQ (open circle); RL (solid circle) © 2004 Oxford University Press

Sikora *et al.* (2007) described the dependence of radio loudness, $R = L_R/L_{opt}$, on the Eddington ratio, $\Lambda = L/L_{Edd}$. Radio-selected AGNs (hosted by elliptical galaxies) can be ~1000 times radio-brighter than optical-selected AGNs (hosted by disc galaxies), but both show a decrease in R with increasing Λ. The Eddington ratio rarely exceeds 0.3 for RL quasars, and the morphological distinction disappears at high-Λ where most elliptical galaxies are radio quiet. It suggests a further dependence on black hole spin for radio loudness.[271] This is consistent with models in which giant elliptical galaxies underwent at least one major merger in the past where the central black hole attains maximum spin upon a twofold increase in its mass.* The mass increase can occur in a relatively short timeframe as a significant amount of cold gas finds its way to the center. Latent signs of mergers/interactions have been found in the available deep images of edge-brightened FR II radio morphologies.[272] The major merger scenario thereby explains the association of radio jet power with factors of galaxy morphology, accretion rate, black hole mass, and spin.

FR II radio galaxies show an interesting anticorrelation with BAL active galaxies. BALs are AGNs having rest wavelength UV spectra of deep, blue-shifted absorption troughs with typical widths of 10,000 km s^{-1}. High-ionization BALs (HiBALs) are associated with strong resonance lines of

---

* See Chapter 4, **Black Hole Rotation (Spin)**.

$C_{IV}\lambda1549$, $Si_{IV}\lambda1397$, $N_V\lambda1240$, $O_{VI}\lambda1035$, and Lyα $\lambda1216$. Low-ionization BALs (LoBALs) tend to a redder continuum with strong $Mg_{II}\lambda2798$ and occasional $Fe_{II}\lambda4570$ lines. Approximately 15% of quasars have BAL features, whereas HiBALs are ~8 times more common than LoBALs. Differences in their emission-line properties from non-BALs (e.g., line ratios, equivalent widths, and levels of ionization) render unlikely any scenario in which BALs are normal quasars viewed edge-on.[273] BALs are typically radio-quiet and are present mostly at accretion rates of $L/L_{Edd} > 10^{-2}$. Gregg *et al.* (2006) examined ~45,000 quasars from the SDSS DR3 catalog to quantify the anticorrelation of radio loudness with BAL quasars. The bias toward higher absorption with lower radio power is clear. FR II morphology exists in ~2.8% of the SDSS quasars, and only five have BAL features in their optical spectra. If FR II and BAL were independent quasar classes, then the SDSS DR3 database should reveal hundreds of quasars in the FR II-BAL subclass. The conclusion is "BALs result from a relatively early phase in the evolution of a quasar during its emergence from a thick shroud or cocoon of dust and gas. Radio emission, particularly the emergence of radio jets to create FR II sources, is frustrated by the obscuring BAL shroud until the quasar can boil away enough of the material through radiation pressure. Observational and theoretical considerations predict that BAL phenomena should preferentially be seen in the early stages of massive black hole mergers, although BALs may also result from minor merger episodes."[274] The suggested evolutionary path* is given as major merger → LoBAL → LoBAL + FR II → (very brief) HiBAL + FR II → non-BAL radio emission.† The broad lines disappear in the latter stages of the evolutionary process due to an expanding RIAF. When the gas ions stop losing energy through radiation, the gas becomes less gravitationally bound to the SMBH and radio outflow is amplified. Trump *et al.* (2011) described changes that occur in the disk as the high-Λ BAL AGN transforms to a low-Λ narrow-line AGN. Figure 6.6 depicts the evolution.[275]

* There are other paths to RL AGNs. Heavy accretion of hot gas from the interstellar medium leads to an FR I. A single infalling star onto a galactic black hole causes a tidal disruption event that creates a radio jet for a period of several months.[276] Such alternate scenarios of RL AGNs do not fall within the scope of EGADS.

† Radio emission does not imply a RL quasar. The majority of non-BALs radiate at $R \ll 10$.

An observed relationship exists in RIAFs (typically at $L/L_{Edd} < 10^{-2}$) between the luminosity of the core radio jet, the black hole mass, and the hard X-ray luminosity. It is given by the equation

$$\log L_R = 0.6 \log L_X + 0.78 \log M_{BH} + 7.33,$$

where the radio luminosity is measured at 5GHz and the X-ray luminosity at 2-10 keV. This important relationship is known as the *fundamental plane of black holes* (FPBH). The analysis of Merloni *et al.* (2006) shows unequivocally that FPBH holds over a large range of distances.[277] The physical origin of the relationship is not fully understood, but its non-linearity betokens an intrinsic characteristic of accreting black holes. Most interestingly, the relationship does *not* hold in radiatively efficient accretion flows. That provides the following curious set of conditions:

- The radio-loud population decreases with redshift
- Radio-loudness is a latter stage of AGN evolution
- The fundamental plane of (RIAF associated) black holes is not a distance artifact

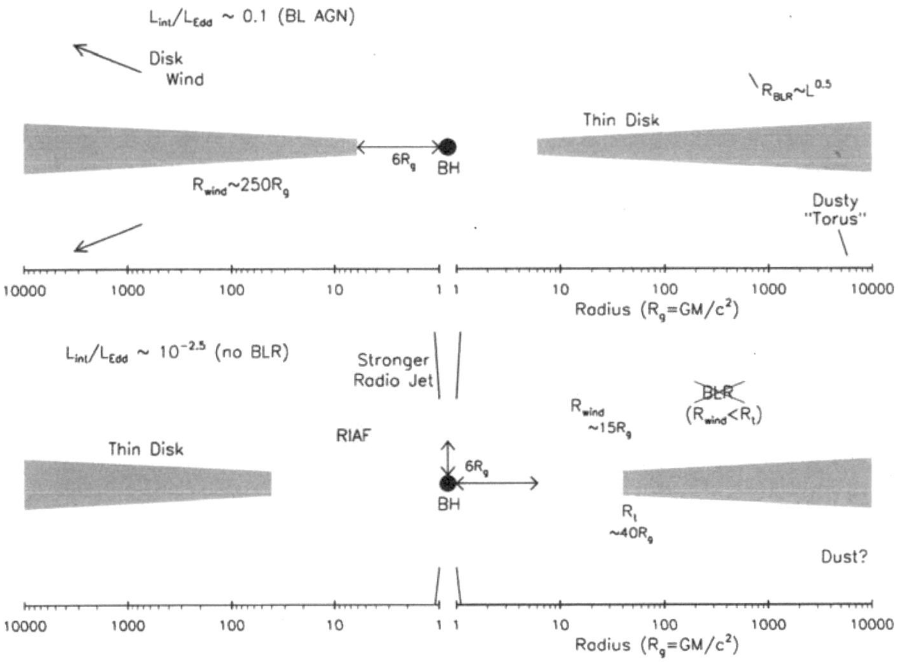

Figure 6.6 – Radio evolution from high-Λ BAL to low-Λ RIAF

The possibility of erroneous distance measurements becomes noteworthy. These conditions would be more self-consistent if it were the case that intrinsic redshift exists during the early stage of AGN (major merger) evolution. Further research of the RL/RQ dichotomy provides opportunity to develop a more consistent picture. Such research must be in a context that has no dependence on source luminosity and no known dependence on radio properties. A PCA of distance-independent parameters provides such a context. The underlying driving factor(s) for all significant correlations to radio-loudness becomes the *prime directive.*

4DE1 – The Four Dimensional Eigenvector Parameter Space

By applying the method of principal component analysis, Boroson, Green and Corbin opened a splendid gateway into the field of AGN structure. Many astrophysicists then followed the 'eigenvector banner' in the crusade to discover properties that differentiated the AGN classes. Two leading astrophysicists who led the quest were Jack Sulentic and Paola Marziani. They and their collaborators began eigenvector work in 1996 and have since produced many important publications. Appendix B chronicles the bounty of their publications.

Sulentic *et al.* (2000c) introduced the optimal three-dimensional correlation space (3DE1) based upon empirical data of low-*z* type-1 BAL AGNs. The three principal correlates assess:

1) FWHM($H\beta_{BC}$)
2) $R(Fe_{II}) = EW(Fe_{II}\lambda4570)/EW(H\beta_{BC})$
3) Γ_{soft}

where Γ_{soft} is the soft X-ray photon index.* The correlation space successfully discriminates between: 1) regular and broad-line quasars; 2) broad-line and narrow-line Seyfert 1 galaxies; and 3) steep- and flat-spectrum broad-line radio galaxies. Two-dimensional projections of 3DE1 are shown in Figure 6.7.[248] Solid symbols represent the RQ sources, while open and crossed circles represent the flat-spectrum and steep-spectrum RL sources. Each diagram shows a clear separation between RQ and RL sources. RL sources show stronger FWHM($H\beta$), weaker $R(Fe_{II})$, and no evidence for

* See Chapter 3, **X-ray Spectrum.**

soft X-ray excess.* RL sources also exhibit ~5 times higher optical continuum luminosities than the RQ sources, even though continuum luminosity is uncorrelated with E1 parameters.† The RQ objects present no evidence for a correlation between 3DE1 coordinates and radio continuum luminosity. Notice the fourth plot in the lower right corner. Unlike $R(Fe_{II})$, $EW(Fe_{II})$ does not correlate as strongly with FWHM(Hβ). The RQ/RL dichotomy diminishes when $EW(Fe_{II})$ substitutes $R(Fe_{II})$.

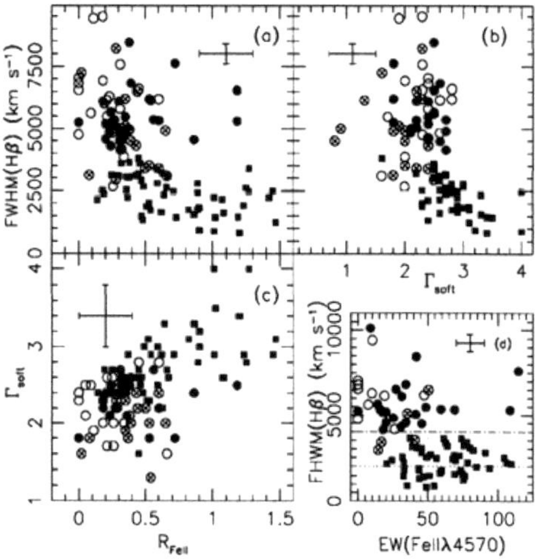

Figure 6.7 – The three correlation planes of 3DE1
RQ (solid);
FR I (open circle);
FR II (crossed)

Marziani *et al.* (2001) took the first step toward the physical interpretation of the E1 eigenvector.[278] Parameters deemed most likely to govern correlations and data point distribution for type-1 quasars in E1 space are FWHM(Hβ_{BC}), $R(Fe_{II})$, Γ_{soft}, and a fourth parameter, $c(½)$, the centroid shift at half maximum of $C_{IV}\lambda1549$. The fourth variable completes the definition of the *Four Dimensional Eigenvector Parameter Space* (4DE1). Introduction of the high ionization line $C_{IV}\lambda1549$ into the E1 space facilitates the important comparison to the low ionization line, Hβ. Both $C_{IV}\lambda1549$ and Hβ are less contaminated by nearby lines and thereby render a clear HIL-LIL distinction. Figure 6.8 shows the extreme correlation of $C_{IV}\lambda1549$ (blue) line shifts with the RQ objects.[279]

Further division of the parameter space along the FWHM(Hβ_{BC}) axis provides a systematic view of the main observational properties. Consider the E1 optical plane (optical components Hβ and Fe_{II}) in the upper left diagram of Figure 6.7. An arbitrary (horizontal) line drawn through the ordinate at 4000 km s^{-1} will define two disjoint populations.

* See Tables 6.1 and 6.2 for PCA correlations.

† Recall that source luminosity was found to be part of the second eigenvector and is therefore not directly correlated with the eigenvector 1 parameters.

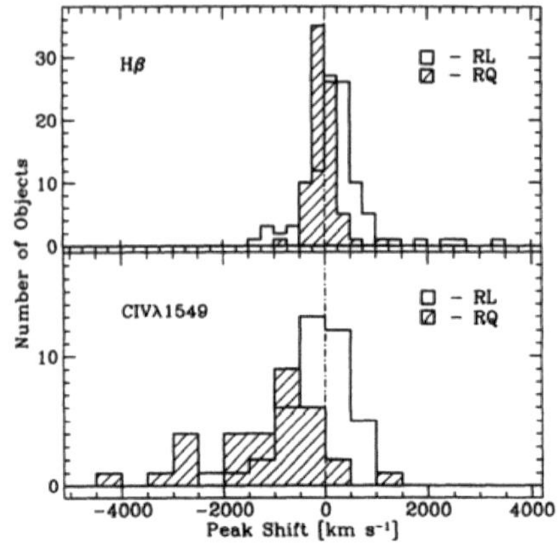

Figure 6.8 – Distribution of rest frame peak shifts for $H\beta_{BC}$ and $C_{IV}\lambda1549_{BC}$ ($z < 0.5$)

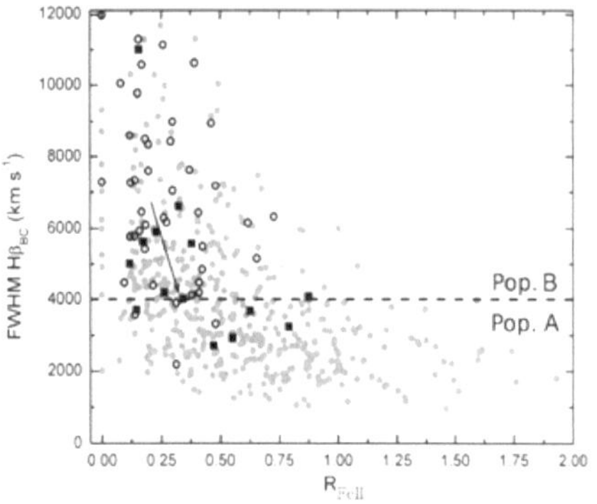

Figure 6.9 – 4DE1 optical plane ($z < 0.7$) RQ (grey dot); FR I (solid square); FR II (open circle)

† **Population A**: FWHM($H\beta_{BC}$) $\leq$ 4000 km s^{-1}, an average of $R(Fe_{II}) \approx 0.7$, significant soft X-ray excess , and $c(½) \approx -800$ km s^{-1}.

† **Population B**: FWHM($H\beta_{BC}$) $\geq$ 4000 km s^{-1}, an average of $R(Fe_{II}) \approx 0.3$, no soft X-ray excess, and no $C_{IV}\lambda1549$ centroid shift.

The differences between population A and B sources reveal a distinction more fundamental than the RQ/RL dichotomy. Comparison between the HIL and LIL in pop-A suggests that they are emitted in disjointed regions.[280] The spectral shape of $H\beta$ is noticeably different between the two populations. Pop-A $H\beta$ shows an unshifted Lorentzian shape, while pop-B $H\beta$ usually requires a double Gaussian model (broad unshifted + very broad redshifted components) whose profiles are indistinguishable between RL and RQ sources.[281] Narrow-line kinematics also changes along the E1 sequence.[282] An increase of $R(Fe_{II})$ exists in pop-A along with systematic decrease in ionization level and increase in electron density. This is most likely due to collision suppression of $H\beta$. Systematic blueshift is not observed in pop-B.[283] Black hole masses are found 1-2 dex larger for pop-B than for pop-A sources.[284]

Figure 6.9 depicts the arbitrarily selected division on the 4DE1 optical plane. Grey dots represent the RQ sources, solid squares are core-dominant RL sources, and open circles are lobe-dominated FR II sources.[285] Three immediate conclusions were reached with the A/B construct:

1) A clear dichotomy exists between RL and RQ AGNs. RL sources are rarely found in the domain of population A, while the RQ sources found in population B rarely exceed FWHM(Hβ_{BC}) = 6000 km s^{-1}.
2) Source orientation cannot account for the E1 correlations, but it does become important as FWHM(Hβ_{BC}) increases for population B.
3) L/L_{Edd} drives the radio-quiet correlation for population A and can be said to define an AGN 'main sequence'. The value $L/L_{Edd} \approx 0.2 \pm 0.1$ corresponds to the boundary between population A and B sources.

Eigenvector 1 (E1) provides a remarkable amount of information. With L/L_{Edd} as its principal component, high and low accretion rates become distinguished on the basis of two disjoint sets of properties. Bifurcation of the E1 further reveals a dichotomy between RL and RQ sources, although an overlap does exist at the corresponding value of FWHM(Hβ) = 5000 ± 1000 km s^{-1}. The RL/RQ dichotomy further sharpens if eigenvector 2 is brought into the picture.

Boroson (2002) constructed PC1 *vs.* PC2 diagrams that reveal diagonal lines of constant black hole mass.[286] PC1 is dominated by the inverse correlation between the strengths of Fe_{II} and [O_{III}]; PC2 is dominated by luminosity and its anticorrelation with the strength of He_{II}. The diagonal line in Figure 6.10 provides a clear demarcation between the RQ (solid) objects and RL (open) objects. The banana shape distribution in the 4DE1 optical plane has been replaced by a clean slice of two-eigenvector pie. Figure 6.11 is a cartoon that Boroson devised to further signify the relevance of M_{BH}. At fixed luminosity, an object that decreases in L/L_{Edd} will increase in mass and transform from RQ to RL. At fixed (high) L/L_{Edd}, an object that increases in luminosity will transform from an NLS1 to a BAL QSO. Luminosities

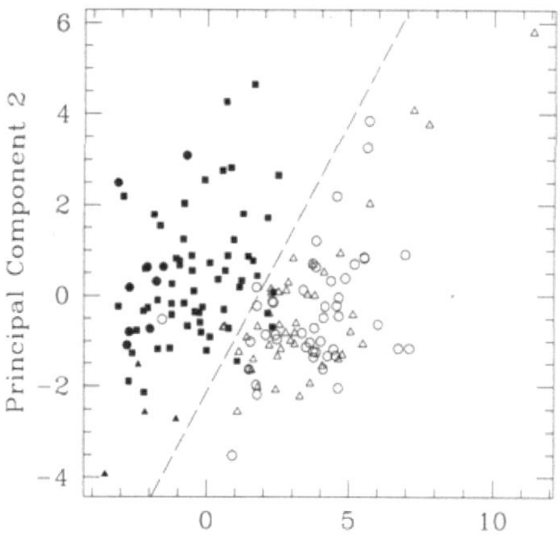

Figure 6.10 – Distribution in PC1-PC2 plane RQ (solid); FR I (triangle); FR II (open circle)

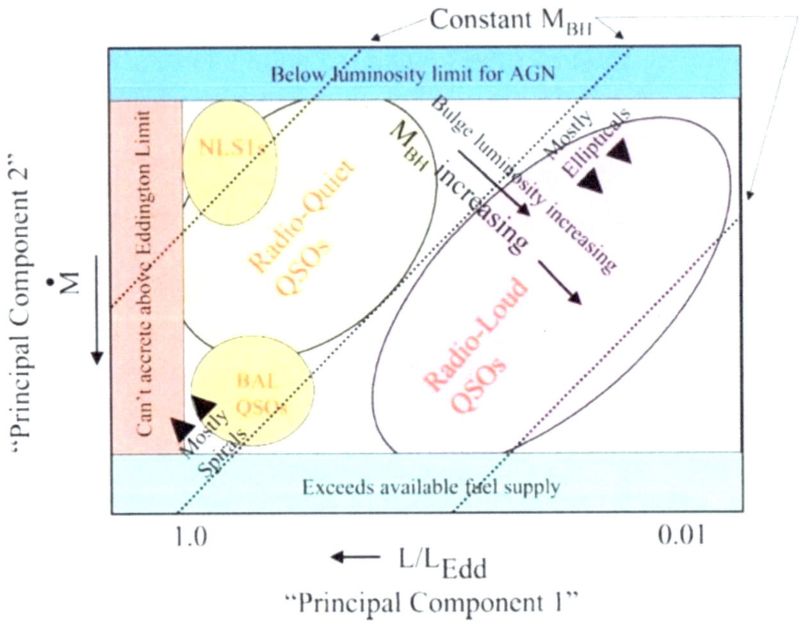

Figure 6.11 – Interpretive diagram of PC1-PC2 plane

less than NLS1 are too faint for detection. Luminosities greater than BAL QSO do not exist for lack of available fuel supply.* Projection of the PC1-PC2 plane onto the L/L_{Edd} axis recreates eigenvector E1 established in 4DE1. It shows an RQ/RL dichotomy and the decrease in M_{BH} against an increase in L/L_{Edd}, but there no longer exists the distinction between NLS1 and BAL QSOs. The distance-independent analysis requires the omission of PC2.† Fortunately, this does not come at the loss of generality. Marziani *et al.* (2003b) showed the influence of black hole mass (and spin) along the 4DE1 AGN main sequence,[287] and Zamfir *et al.* (2008) showed that 4DE1 parameters have no obvious dependence on source luminosity.[288]

Quasar Evolution at Low Redshift

The current opinion is that L/L_{Edd} and M_{BH} serve as arrows of time in the life cycle of a quasar. Analysis of infrared (IR) color diagrams of low-z ultraluminous IR galaxies (ULIRGs) reveals an AGN evolutionary sequence,

* Boroson found no indication that orientation is a determining factor in the observable characteristics of AGNs.

† The luminosity factor also presents a high risk of selection bias.

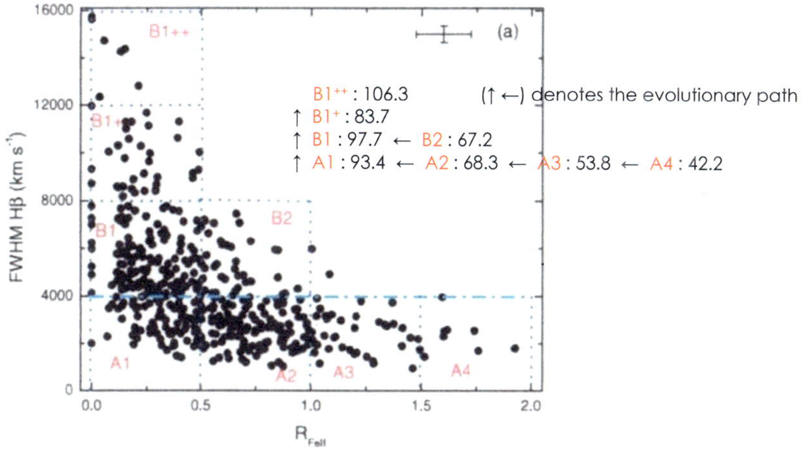

Figure 6.12 – 4DE1 optical plane

independent of luminosity, that commences with starbursting in a strong interacting system of very high-density gas (Krongold *et al.*, 2002)[289]. Through multiple duty cycles, the accretion rate of the once enshrouded object (in population A) gradually diminishes as black hole mass and spin attain maximum values (in population B).* It thus becomes possible to identify the phase of quasar evolution by spectral analysis of its 4DE1 parameters, all of which are correlated to L/L_{Edd}. Consider the spectral evolution of (Hβ)λ4861. Zamfir *et al.* (2010) provided a detailed characterization of Hβ emission lines from ~470 SDSS DR5 low-*z* quasars.[291] The 4DE1 optical plane is subdivided into eight bins of ΔFWHM(Hβ) = 4000 km s^{-1} and $\Delta R_{\mathrm{FeII}} = 0.5$ (see Figure 6.12). Composite properties in each bin exhibit marked differences. The median bin values of EW(Hβ)λ4861 (overlaid on the figure) show clear evolution:

Figures 6.13 and 6.14 show the composite spectra for bins A2 and B1 in the optical range of 4600-5050 angstroms. Easily observed is the doubling of FWHM(Hβ) from bin A2 to bin B1 (2690 km s^{-1} to 5368 km s^{-1}). Also quite evident is the much larger value for peak [$\mathrm{O_{III}}$]λ5007 in B1 predicted by the principal component analysis. Such distinctions are not limited to the optical emission lines. UV spectral diagnosis reveals low equivalent widths in *all* of the strongest broad emission lines ($\mathrm{C_{IV}}$λ1549 and the Balmer UV lines), and clearly distinguishes the density and ioniza-

* Some researchers further suggest that type-2 precedes type-1 AGNs in the evolutionary sequence.[289] [290]

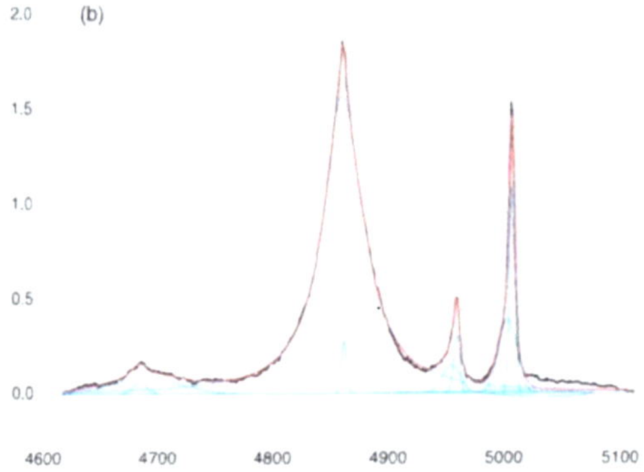

Figure 6.13 – Bin A2 composite
© 2010 Oxford University Press

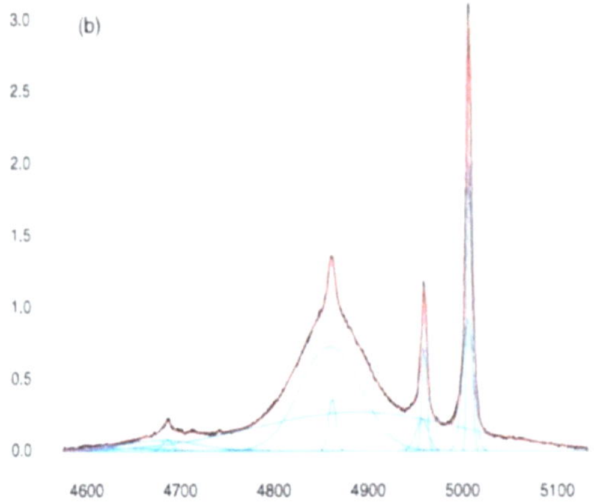

Figure 6.14 – Bin B1 composite
© 2010 Oxford University Press

tion levels at the extreme ends of the 4DE1 sequence.[292] Bensch *et al.* (2015) demonstrated that significant changes exist in soft X-ray excess between populations A and B.[293] Neither luminosity nor black hole mass drives the observed spectral diversity.[294]

A large corpus of information has been stored in AGN repositories throughout the last several decades. PCA identifies the most important properties to study within all that stored information. The 4DE1 formalism systematizes those properties in a manner that discriminates sources by $L/L_{\rm Edd}$, a documented arrow of time. Property changes are observed in the optical, UV and X-ray emission lines. Fraix-Burnet *et al.* (2017) provided further support of the evolutionary process by means of cladistic analysis of all 4DE1 parameters.[295] There now exists a reasonable picture of low-*z* quasar evolution. Furthermore, any faulty distance measurements are not likely to change the current representation. The 4DE1 parameters are distance-independent, and redshift measurements are fairly well established in the $z < 0.5$ universe.*

Quasar Evolution at Intermediate Redshift

Both CH and EGADS must offer some explanation of quasar evolution beyond low-z. The luminosity and density functions are particularly difficult for CH to explain. EGADS offers the more straightforward explana-

* Chapter 2 provides the arguments in favor of the cosmological hypothesis for $z < 0.5$.

tion on the basis of intrinsic redshift. A rationale for quasar spectral evolution within the 4DE1 framework is given by EGADS theoretical prediction #7 (a, b, c).*

7a. *The ratio of population A to population B objects will increase with redshift.* – Higher intrinsic redshift will move a source from pop-B to pop-A.
7b. *At **higher** redshift, 4DE1 parameter space will evolve with luminosity.* – Intrinsic redshift will cause overestimates in luminosity as well as modify the 4DE1 spectral components.
7c. *At **higher** redshift, 4DE1 parameter space will evolve more significantly in population A than population B.* – Large intrinsic redshift is found most often in pop-A sources.

The demarcation between populations A and B is based on FWHM(Hβ) = 4000 km s^{-1}. The line provides useful distinctions between objects at low-z. This no longer holds at higher luminosity and redshift. The minimum FWHM of low-ionization lines increases monotonically with luminosity,[296] and no source is found below FWHM(Hβ) = 3500 km s^{-1} at $z > 3$.[297] A possible explanation for the increase in FWHM is larger M_{BH}, but the question remains how the high-z quasars evolve to larger mass (and luminosity) than their low-z counterparts.

Analysis of spectral differences between the two populations at higher redshift requires an adjustment to the binning found in Figure 6.12. It is not at all clear how this should be done. EGADS theoretical predictions #7a and #7b imply that bin lines will become blurred. Spectral composites based on luminosity variance are available, but they are inadequate for distance-independent analysis. It is necessary to analyze the *distance-independent* LIL, HIL and X-ray properties in *each population per unit redshift.* This is beyond the means of existing technology. Although quasars are now easily detected all the way to $z = 4$, most emissions are too faint for proper analysis. A weaker statement of prediction #7c becomes necessary for current scrutiny.

* See Chapter 5 for the first six EGADS theoretical predictions.

7c. *Component composites of 4DE1 parameters will change noticeably by redshift in bins A2 and A3; No significant changes by redshift will occur in bins B1 and B1+.*

LIL properties - Analysis of the 4DE1 optical plane at intermediate redshift requires high S/N infrared spectroscopy to observe the (weakened) emissions that have moved out of the optical band. Sufficient numbers of IR spectra first became available in 2009 from VLT-ISAAC (Infrared Spectrometer And Array Camera) at the European Southern Observatory. Sulentic, Marziani *et al.* combined ISAAC objects ($0.9 < z < 3.0$) with previous data to analyze LIL profiles. Their work was presented in a series of three papers that concentrated on the Hβ broad-line region. In the first paper (Sulentic *et al.*, 2004) 17 ISAAC spectra show no strong evolution with redshift, but NLS1 sources with $M_B = -28$ reach as much as 2000 km s^{-1} larger FWHM(Hβ) than the sources with $M_B = -22$. There also is evidence that population A sources increase with redshift. [298] The second paper (Sulentic *et al.*, 2006) deduces possible twofold M_{BH} overestimations in the intermediate redshift sources.* [146] FWHM(Hβ), the accepted estimator of virial velocity, consists of the broad component FWHM(Hβ_{BC}) and the very broad component FWHM(Hβ_{VBC}). The redshift of FWHM(Hβ_{VBC}) suggests that it is affected by broadening other than that arising from the virial gas. Sun and Shen (2015) also asserted that the vertical spread in FWHM(Hβ) at fixed R_{FeII} does not track the spread in true M_{BH}. [299] Possible causes for the variance in FWHM(Hβ_{VBC}) are black hole mass, spin and orientation. Whatever the cause, the phenomenon seems to occur during the latter stage of quasar evolution and is restricted to population B. The third paper (Marziani *et al.*, 2009) extends the examination of LIL profiles to 53 ISAAC objects. Distributions found in the profile parameters are essentially the same at all luminosities, but almost half are displaced towards higher values of R_{FeII} and FWHM(Hβ). The systematic increase in minimum FWHM(Hβ) with bolometric luminosity (L_{bol}) appears only in population A sources. No such trend occurs in population B, where sources remain at a lower Eddington ratio. [300]

* This would compound any M_{BH} overestimations made by faulty luminosity measurements. See Chapter 5, **Black Hole Mass Estimations**.

HIL properties - The most striking difference between the low-*z* and ISAAC samples occurs in the high ionization $[O_{III}]\lambda5007$ narrow line. Approximately 80% of the ISAAC lines are weak (several angstroms or less), and about 40% are blueshifted by more than 250 km s^{-1}.[301] Shen and Ho (2014) used 20,000 broad-line quasars ($0 < z < 0.9$) from SDSS DR7 to illustrate the diversity of the $[O_{III}]$ emissions in the optical plane. Although the bin sizes are at variance with the 4DE1 optical plane, binned data are in the same relative positions as found in Figure 6.12.[302] Figure 6.15 shows vividly how $[O_{III}]$ emissions (low- and high-*z* quasars combined) are weaker in A3 than A2. It also clearly evidences no change in line strength from B1 to B1+. Figure 6.16 shows $[O_{III}]$ composite spectra of each bin. There is no evolution in spectra along the B1 column. The evolution along the horizontal rows, however, includes weaker emissions and growing blueshifted wings.* Another noteworthy item is that every bin exhibits weaker emissions as a function of luminosity. This demonstrates the ubiquity of the Baldwin effect for $[O_{III}]$. Bachev *et al.* (2004) showed a similar effect in the

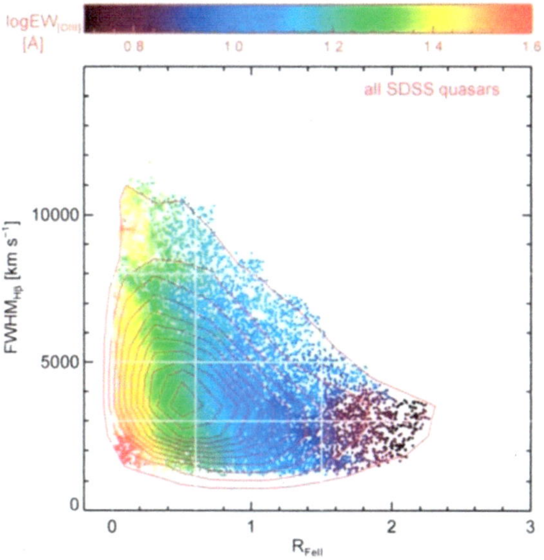

Figure 6.15 – Quasar distribution in optical plane
© 2014 Springer Nature

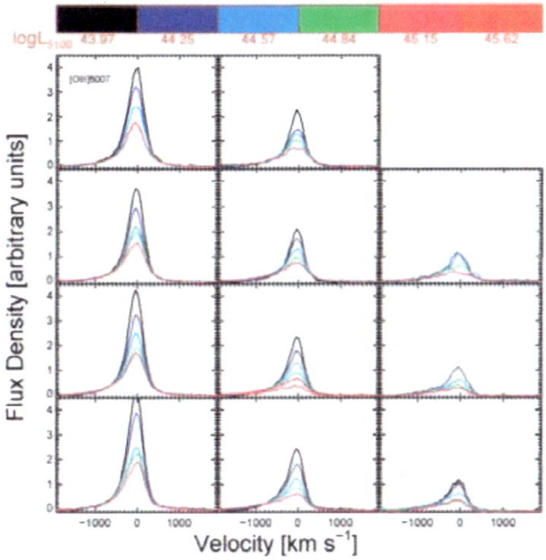

Figure 6.16 – [OIII] profiles in optical planes
© 2014 Springer Nature

* Some of the profile diversity is dampened in Figure 6.16 due to nonuniform binning.

composite spectra of other strong high-ionization lines (Lyα, $C_{IV}\lambda1549$, $Si_{IV}\lambda1398$, and $C_{III}\lambda1909$).[303]

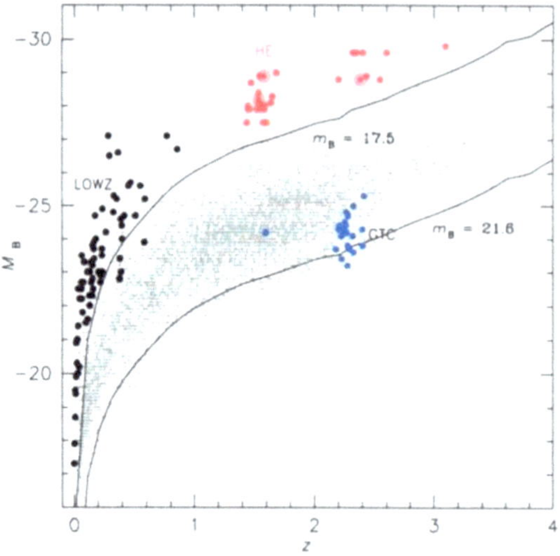

Figure 6.17 – Location of objects in the low-z, GTC and HE samples

The HIL blue-asymmetric profiles increase with redshift and occur in both the BLR and NLR.[304][305] The underlying cause remains in question. Comparison between low and high redshift is best made with the $C_{IV}\lambda1549$ profile. It is the safest feature for a BLR diagnostic beyond $z \sim 1.4$ because it does not blend heavily with other strong lines. Two recent papers by Sulentic, Marziani *et al.* examine C_{IV} profiles in detail in order to offer the best possible evidence of the relative quasar behavior in populations A and B, at low- and high-*z*, and with low and high luminosity. Sulentic *et al.* (2014b) compared 22 GTC (Gran Telescopio Canarias) objects at redshift $2.3 < z < 2.5$ and absolute magnitude $-25 < M_B < -23$ against local luminosity analogs acquired from the *HST*-FOS archive. This effectively compares some of the lowest luminosity high-*z* sources against some of the highest luminosity low-*z* sources. The two samples have a similar population A/B dichotomy with spectral differences preserved, but small discrepancies exist. Populations A and B of the high-*z* GTC sample show FWHM ($C_{IV}\lambda1549$) with averages wider by 28% and 8% over the corresponding populations of the low-*z* FOS control sample. The GTC sample shows lower chemical abundances in the line-emitting gas.[306] This is at some noticeable variance with the known correlation of metallicity to luminosity. The EGADS influence on the GTC sample is difficult to assess. It is not possible to know whether any particular high-*z* quasar contains intrinsic redshift, and the restriction to low luminosity sources may cause selection bias against any sources that do. The increase in FWHM ($C_{IV}\lambda1549$) and decrease in metallicity do, at least, tend to favor some possible intrinsic redshift.

Spectra of 52 higher luminosity objects at high-z were obtained from the Hamburg ESO (HE) VLT-ISAAC NIR survey. The mostly radio-quiet sample ($0.9 < z < 3.1$; luminosity $\log L > 47.4$) was compared against a control sample of 130 RQ low-z *HST*-FOS $C_{IV}\lambda1549$ observations. If intrinsic redshift exists, the HE sample is the most likely place to find it. Figure 6.17 shows the relative positions of the low-z (black), GTC (blue) and HE (red) objects in the absolute magnitude *vs.* redshift plane. Note that low-z quasars do not exist with luminosities similar to those in the HE sample. Sulentic *et al.* (2017) found significant differences in C_{IV} line profiles between the low-z and HE samples.[307] Populations A and B of the HE sample show FWHM ($C_{IV}\lambda1549$) with averages wider by 50% and 22% over the corresponding populations of the low-z sample. Most remarkable is the fact that HE C_{IV} $c(\frac{1}{2})$ blueshifts exceed 2000 km s^{-1} in almost all cases. Such large shifts are extremely rare at low-z. It seems that $c(\frac{1}{2})$ tracks L/L_{Edd} at low redshift, but it also tracks luminosity at high redshift. This is difficult to explain with the wind outflow model for blueshift. Figure 6.18 shows the $c(\frac{1}{2})$ amplitude of HE objects in the 4DE1 optical plane. Population A and B sources are represented by circles and squares respectively.

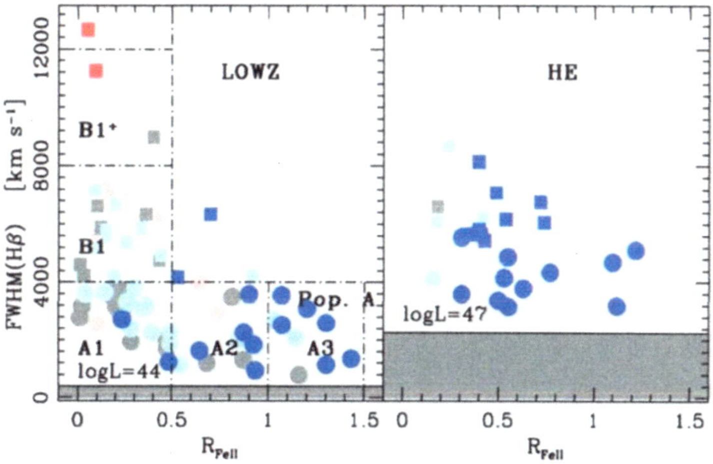

Figure 6.18 – C_{IV} $c(\frac{1}{2})$ in the 4DE1 optical plane
© 2017 EDP Sciences

Blue: $c(\frac{1}{2})$ blueshift exceeds –1000 km s^{-1}
Aqua: $c(\frac{1}{2})$ blueshift of –200 to –1000 km s^{-1}
Grey: $|c(\frac{1}{2})| < 200$ km s^{-1}
Red/pink: $c(\frac{1}{2})$ strong to moderate redshift

Note how the demarcation between A and B sources has moved upward along the FWHM(Hβ) axis. The distinction between extreme pop-A and extreme pop-B has diminished in the HE sample. It will be important to eventually learn whether this is a selection bias towards high-η (high accretion efficiency) sources. In any case, it is clear that $c(½)$ becomes stronger at higher redshift in sources with higher luminosity.

X-ray properties –There is yet no indication of any significant evolution in X-ray properties of quasars over the bulk of cosmic time.* Whatever discovery is made in its evolution must then reveal something noteworthy. Soft X-ray excess, Γ_{soft}, is the important parameter in population A sources; EGADS predicts Γ_{soft} evolution with redshift. Past studies of X-ray properties have been severely constricted by small samples and low quality data. This has changed with the advent of *XMM-Newton,* which now provides quality spectra in the range of 0.5-10 keV ($0.5 < \Gamma_{soft} < 2$ and $2 < \Gamma_{hard} < 10$). Unfortunately, there remains little sampling of soft X-rays at higher redshift. Current analysis of X-ray evolution must rest mainly on Γ_{hard}, which also happens to be the best method of selecting all AGN types. Risaliti *et al.* (2009) demonstrated a significant correlation between Γ_{hard} and L/L_{Edd} in an SDSS sample of ~400 (RL and BAL excluded) quasars in the redshift range $0.1 < z < 4.5$.[308] The increase of L/L_{Edd} with redshift gives some hint of evolution in Γ_{hard}. Brightman *et al.* (2009) reinforced this finding with measurements of Γ_{hard} from COSMOS (Cosmic Evolution Survey) and E-CDF-S (Chandra Deep Field-South) broad-lined AGNs ($0.5 < z < 2.1$). The correlation,

$$\Gamma_{hard} = \left(0.32 \pm 0.05\right)\log_{10}\Lambda_{Edd} + \left(2.27 \pm 0.06\right),$$

shows non-negligible evolution when the entire sample is divided into two redshift bins ($0.5 < z < 1.2$ and $1.2 < z < 2.1$).[309] Similar measurements of Γ_{soft} are necessary.

The evolution of the 4DE1 parameter space with redshift is not in doubt, but there remains a dilemma. The steep X-ray spectra of high L/L_{Edd} pop-A sources suggest slim ADAF disks, but that begs further explanation for the strong Fe_{II} line profiles and the strong high ionization winds.[310] Examination of the associated host morphologies improves the picture. Brightman *et al.* (2010) found evidence for two important factors in the

* See Chapter 3, **X-Ray Spectrum**.

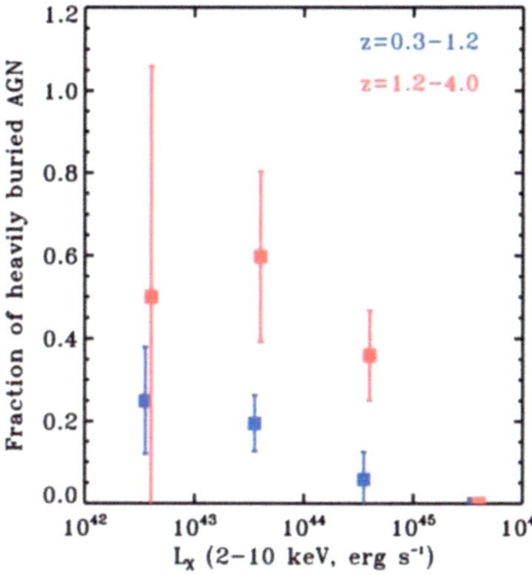

Figure 6.19 – CT-AGN vs. L_X in two z-bins
© 2014 Oxford University Press

CDF-S subset of Compton thick X-ray sources ($0.1 < z < 4$).* First, The covering factor of the obscuring material *decreases* with L_X. This is consistent with predictions that the torus recedes during quasar evolution. Second, the covering factor *increases* with redshift. Figure 6.19 plots the fraction of heavily covered AGNs as a function of L_X for two redshift bins.[311] No consensus exists for the larger covering factor at higher redshift. A contributing factor could be compact nuclear starbursts within the host galaxy following a merger / interaction event. Trakhtenbrot *et al.* (2017) used ALMA millimeter/sub-millimeter data of high-*z* quasar host observations to study the early SMBH growth and star formation (SF) driven by major merger.

> …All this supports a general idea that both processes (SF and AGN) are fed by a common reservoir of cold gas that collapses, forms stars, and (eventually) reaches the central region of the host galaxy to be accreted by the SMB. …Moreover, there is strong evidence that the occurrence rate of mergers among AGN increases with increasing AGN luminosity and redshift.[312]

By analysis of their host morphologies, Kocevski *et al.* (2015) showed that Compton thick AGNs are three times more likely to be associated with a merger/interaction signature.[313] This further supports the scenario that slim disks are found at an early phase of quasar evolution. Heavily obscured CT-AGN sources and the higher incidence of extremely luminous pop-A are thus distinguishing factors of quasar evolution at intermediate redshift. They are also additional factors that support EGADS theoretical prediction #7.

---

* A Compton thick source is one with column density of $N_H > 10^{24}$ cm^{-2}.

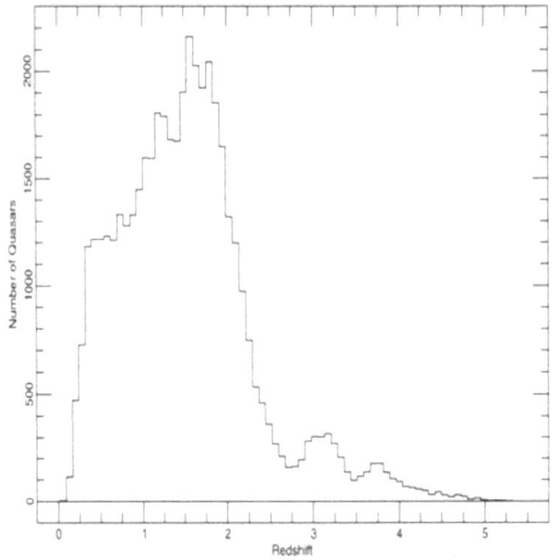

Figure 6.20 – Histogram of SDSS DR3 quasars

Quasar Evolution Over Cosmic Time

The evolution of quasars presents a profound mystery. The salient enigma is their density and luminosity distributions over cosmic time. Quasars were on average 100 times more luminous ten billion years ago, and there were far more of them. Nothing comparable is required to explain galaxy evolution over the same time frame. At high-*z* the space density becomes minuscule by comparison, while luminosity grows to inexplicable magnitudes.* Figure 6.20 shows the histogram of 46,420 objects from the SDSS DR3 catalog.[314] Although high-*z* quasars are accessible to telescopes, virtually nothing is revealed about quasar evolution during the early epochs of diminished space density. A tenable evolutionary scenario now exists for the last ten billion years ($0 < z < 1.36$) from the acquired low- and intermediate-*z* data, but it unfolds a further mystery within a mystery.

The low- and intermediate-*z* evolutionary epochs are in striking imbalance given the quasar life cycle of major merger → non-BAL radio emission.† Low-*z* shows little incidence of extreme BAL, Compton thick systems that purportedly exist in the early phase after a major merger.‡ Intermediate-*z* has a dearth of radio-loud AGNs that should exist in the end phase. The latter is particularly difficult to explain. Jiang *et al.* (2007) determined the radio-loud fraction (RLF) of more than 30,000 quasars as a function of redshift ($0 < z < 5$).[315] Figure 6.21 is representative of the data, which clearly show that the RLF strongly decreases with increasing redshift at any constant luminosity. This places a major constraint on the radio

* See Chapter 3, **Evolution**, for QSO evolution that defies explanation.

† See previous section, **Radio Loudness and Evolution**.

‡ The small number of broad-line AGN at low X-ray luminosities is known as the *Steffen Effect*.

emission mechanism. It also exacerbates our understanding of quasar cosmic evolution. Is there a better picture?

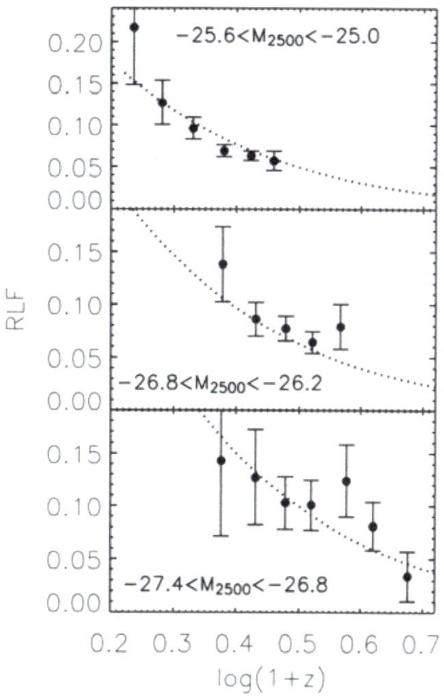

Figure 6.21 – Radio Loud Function vs. z

Consider Figure 6.22, a cartoon representation of accretion disk evolution during the life cycle of the quasar. The vertical axis indicates the time τ since quasar birth. The early disk structure can be α-mode, slim, thin or β-mode.* Each structure is bounded by a range of values of the undefined parameter, Z, along the horizontal axis. As time progresses, L/L_{Edd} dwindles and all structures ultimately revert into a radiatively inefficient accretion flow.

The upper right corner of the τ-Z plane contains a forbidden zone where thick/slim disks cannot exist long after creation. The Z parameter represents the unknown in quasar evolution. Two possibilities are proposed. First, there are indications that Z may be proportional to z_c. The low L/L_{Edd}, radiatively inefficient LINERs prevail at low-z, the thin disks dominate the intermediate-z space, and the slim/thick disks prefer creation at some higher redshift. This scenario leaves the accretion rate as the primary factor to influence the change in disk structure. The last four chapters of this book, however, reveal the shortcomings of such a pro-

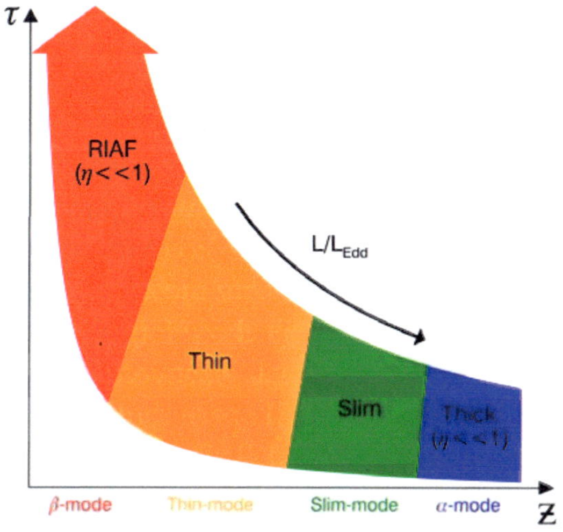

Figure 6.22 – Evolution of accretion disk structure

* See Chapter 5, **Accretion Disk Geometry**.

posal. Second, consider that $\mathcal{Z}$ represents gravitational potential. The disk structure is now a function of both gravity *and* the accretion rate. α-mode disks exist in extreme (EGADS) gravity with super-Eddington accretion. Over time the accretion rate diminishes, the SMBH becomes the sole gravitational factor, and the disk structure ultimately morphs into β-mode. In this latter case, $z_{tot} = z_c + z_i$, where z_i is the gravitational redshift. Other consequences that follow from the gravity proposal for $\mathcal{Z}$ are:

- Thick disks are infrequent at low-*z* where $z_i = 0$.
- RL quasars are infrequent at intermediate-*z* where $z_i > 0$.
- The quasar space density peaks where both $z_i > 0$ and $z_c > 0$.
- The quasar space density falls dramatically as z_i approaches an asymptotic limit.
- Luminosity and distance are overestimated where $z_i > 0$.
- Large HIL blueshifts occur where $z_i > 0$.
- Pop-A spectra evolve with z_i.

These factors eradicate many dilemmata. Technological limitations preclude any strong support of EGADS theoretical prediction #7, but a simple application of Occam's razor will prefer EGADS over the cosmological redshift hypothesis. On that basis alone it must be the case that extreme gravitational redshift, though unsubstantiated, is definitively plausible.

Summary

The spectra of active galactic nuclei offer a wide variety of variables for study. *Principal component analysis* provides the best means to identify the properties that are apt to provide meaningful information. Exclusion of the distance-dependent parameters helps ensure the accuracy of the analysis. Parameters that correlate to radio-loudness are most important. Jack Sulentic, Paola Marziani and their collaborators chose four PCA parameters for construction of the four-dimensional eigenvector space (4DE1). Any two-dimensional projection of the 4DE1 space shows a clear distinction between two source populations, A and B. There emerges a RL/RQ (radio-loud/radio-quiet) dichotomy, as most RL sources appear in population B. Other research suggests that low-*z* RL objects appear as an end phase of quasar evolution. Lack of RL objects at intermediate-*z* thus calls into question the cosmic evolution of quasars. Extreme gravity helps explain the evolution of disk structure. It also eradicates many of the unexplained difficulties associated with the cosmological redshift interpretation. The best available evidence supports the proposition that certain factors in the gravitational redshift of active galactic nuclei have been largely underestimated.

Epilogue

The dedicated scientist would devote a lifetime of effort for the opportunity to unlock only one of the many secrets of our vast Universe. That opportunity comes after a great deal of time and effort has been expended collecting all available information. Time becomes the enemy of the enthusiastic scientist who finds the sheer volume of research prohibitive. Theory must then substitute for fact, and the *best* theory does not always imply a *good* theory. A certain degree of bias becomes inevitable.

I had no idea where my research would lead when I began. I was motivated to understand why solid evidence provided by world-class scientists was dismissed in some cases. Geoffrey Burbidge offered convincing statistical evidence for the close association of QSO-galaxy pairs. Why was it dismissed? For many years William Tifft gathered hard evidence for galaxy redshift quantization. Why was it ignored? More recently, William Gaskell showed that HIL blueshifts are inbound flows. Why is this rejected without any contradictory observations? My knowledge of these matters was limited, and I would start afresh to formulate my own opinion on these matters. After years of research, it is only natural that I would develop a favored way of thinking. It is fair to state that some degree of bias has become associated with my newfound perspective.

Not long before the completion of Chapter 6, my adviser asked me whether I now believed in the hypothesis I had proposed. I rather surprised her with the response: "Not really." I impressed upon her that I believed in the *plausibility* of EGADS to stand as a solid argument against those who too easily accept the standard model and dismiss the works of Burbidge, Tifft, Gaskell, *etc.* The extreme gravitational field seems very probable, but Variable Mass, Time Dependent Decay, and other viable theories must also remain plausible explanations for intrinsic redshift. The relative influence of dark matter and $\mathcal{N}$-body gravitation in the early life cycle of some AGNs is worthy of serious consideration. We should also entertain the possibility that dark matter may be nothing more than slowly cooling white dwarfs, neutron stars, black holes, and cold gas. When everything collapses at relativistic speeds during a merger of dark halos, extreme $\mathcal{N}$-body gravitation creates a different world, enshrouded and radiatively weakened.

So much remains unknown about the most wondrous objects in the Universe. A more developed theory must come from such fields as magnetohydrodynamics, quantum gravity, or other new physics. We may yet discover that QSOs are the pockets of creation imagined by quasi-steady state cosmologists. Fred Hoyle would not see much bias in my vision.

Appendix A – Figures and Tables

Figures and tables reproduced in this book come from many different journals, texts and Internet files. Licenses for reuse have been obtained from Oxford University, Cambridge University, American Institute of Physics, *Annual Review of Astronomy & Astrophysics, Nature, Science, New Astronomy Reviews, Physical Review, Physica Scripta, International Journal of Theoretical Physics, Indian Academy of Sciences,* and *Chinese Journal of Astronomy and Astrophysics.* In certain cases where formal licenses are not required, the individual providing the permission is mentioned. Government images in the public domain are credited in the usual fashion. Images from *Astrophysical Journal* and *Astronomical Journal* are reproduced with permissions from the appropriate authors.

Numbers between [1] and [315] listed in column two are the references that appear in the text. See **References** for author and title information.

Figures

Figure		Source / Credit
Cover		/credit: European Southern Observatory/ https://www.eso.org/public/images/eso0903a/
Figure 1.1	[7]	<http://www.helsinki.fi/~hkurkisu/cosmology/Cosmo4.pdf>, Fig 11 */reproduced with permission of Hannu Kurkisu/*
Figure 1.2		/credit: NASA/STScI - Chandra X-ray Observatory/
Figure 2.1	[316]	*Relativity, Gravitation and Cosmology* (Oxford Univ. Press), Fig 7.3 /reuse license acquired from Oxford Publishing Limited/
Figure 2.2		/credit: ESA/Hubble & NASA/
Figure 2.3	[18]	*Astrophysical Journal Letters*, 777:L17, 2013 November 1, Fig 1 /reproduced with permission of Arjen van der Wel/
Figure 2.4		<pages.astronomy.ua.edu/keel/agn/forest.html> /credit: HST/FOS and Keck I HIRES/

Figure		Source / Credit
Figure 2.5	[26]	*Astrophysical Journal*, 513:576-591, 1999 March 10, Fig 1 /reproduced with permission of Gordon Richards/
Figure 3.1	[38]	*Quasars and Active Galactic Nuclei* (Cambridge Univ. Press), Fig 7.3 /reproduced with permission of Cambridge University Press/
Figure 3.2	[47]	*Astrophysical Journal*, 565:50-62, 2002 January 20, Fig 2 /reproduced with permission of Anca Constantin/
Figure 3.3	[54]	*Astronomical Journal*, 129:2519-2530, 2005 June, Fig 6 /reproduced with permission of Cristian Vignali/
Figure 3.4	[55]	*ASP Conference Series*, Volume 373, 2007, Fig 4 /reproduced with permission of Dr. Joseph Jensen, ASPCS, editor/
Figure 3.5	[60]	*Astrophysical Journal*, 279:19-39, 1984 April 1, Fig 3 /Philipp Kronberg work reproduced with permission of Gary Welter/
Figure 3.6	[63]	*Monthly Notices of the Royal Astronomical Society*, 405, 1940-1946 (2010), Fig 2*a* /reuse license acquired from Oxford University Press/
Figure 3.7	[63]	*Monthly Notices of the Royal Astronomical Society*, 405, 1940-1946 (2010), Fig 5*a* /reuse license acquired from Oxford University Press/
Figure 3.8	[63]	*Monthly Notices of the Royal Astronomical Society*, 405, 1940-1946 (2010), Fig 2*b* /reuse license acquired from Oxford University Press/
Figure 3.9	[70]	Clark, Stuart, *Redshift* (University of Hertfordshire Press, 1997), Chapter 5. /reproduction by Daniel Mayes/
Figure 3.10	[76]	*Astronomical Journal*, 146:120, 2013 November, Fig 13 /reproduced with permission of M.L. Lister/
Figure 3.11	[317]	*Astronomy & Astrophysics,* 18, 455-463 (1997), Fig 2 /© EDP Sciences/
Figure 3.12	[317]	*Astronomy & Astrophysics,* 18, 455-463 (1997), Fig 4 /© EDP Sciences/
Figure 3.13	[91]	*ASP Conference Series*, Volume 51, 1993, Fig 1 /reproduced with permission of Dr. Joseph Jensen, ASPCS, editor/
Figure 3.14	[99]	*Astrophysical Journal*, 616:738-744, 2004 December, Fig 1 /reproduced with permission of Morley Bell/
Figure 3.15		/credit: D. Strange, Worth Hill Observatory, Dorset/
Figure 3.16	[110]	*Astronomy & Astrophysics*, 390, L15-L18 (2002), Fig 1*a* /© EDP Sciences/
Figure 3.17	[111]	*Astrophysical Journal*, 620:88-94, 2005 February 10, Fig 1 /reproduced with permission of Stefano Zibetti/
Figure 3.18	[112]	*Astrophysical Journal*, 500:596-598, 1998 June 20, Fig 2 /reproduced with permission of Yaoquan Chu/
Figure 3.19	[119]	*Astrophysical Journal Supplement Series*, 74:675-730, 1990, Fig 3*a* /reproduced with permission of Jayant Narlikar/
Figure 3.20	[119]	*Astrophysical Journal Supplement Series*, 74:675-730, 1990, Fig 3*c* /reproduced with permission of Jayant Narlikar/
Three Pillars of CH		/produced by Robert Gnaski Burton/

Figure		Source / Credit
Figure 4.1	[12]	*Active Galactic Nuclei* (Cambridge University Press, 1997), Fig 1.9 /reuse license acquired from Cambridge University Press/
Figure 4.2	[145]	*Astronomy & Astrophysics*, 407, 461-472 (2003), Fig 1 /© EDP Sciences/
Figure 4.3	[149]	*Astronomy & Astrophysics*, 457, 485-492 (2006), Fig 1 /© EDP Sciences/
Figure 4.4	[149]	*Astronomy & Astrophysics*, 457, 485-492 (2006), Fig 3 /© EDP Sciences/
Figure 4.5	[149]	*Astronomy & Astrophysics*, 457, 485-492 (2006), Fig 4 /© EDP Sciences/
Figure 4.6	[150]	*International Journal Of Theoretical Physics*, July 2015, Volume 54, Issue 7, pp 2398-2417, Fig 2 /reuse license acquired from Springer Nature/
Figure 4.7	[152]	*Astrophysical Journal*, 620:69-77, 2005 February 10, Fig 5 /reproduced with permission of Piero Madau/
Figure 4.8	[152]	*Astrophysical Journal*, 620:69-77, 2005 February 10, Fig 8 /reproduced with permission of Piero Madau/
Figure 4.9	[154]	*Astrophysical Journal*, 767:37-50, 2013 April 10, Fig 8 /reproduced with permission of Umberto Maio/
Figure 4.10	[155]	*Astrophysical Journal*, 775:94-107, 2013 October 1, Fig 9 /reproduced with permission of Marek Sikora/
Figure 4.11	[160]	*Astronomical Journal*, 134:1150-1161, 2007 September, Fig 1 /reproduced with permission of Xiaohui Fan/
Figure 4.12	[168]	*Physical Review D*, 75:024034, 26 January 2007, Fig 2 /reuse license acquired from American Physical Society/
Figure 4.13	[170]	*Physical Review Letters*, 99:181101, 29 October 2007, Fig 1 /reuse license acquired from American Physical Society/
Figure 4.14	[171]	*Astrophysical Journal*, 238:471-487, 1980 June 1, Fig 6 /reproduced with permission of W. Kent Ford Jr./
Figure 4.15	[177]	*Astrophysical Journal*, 648:L109-L113, 2006 September 10, Fig 1 /reproduced with permission of Douglas Clowe/
Figure 4.16	[192]	*Monthly Notices of the Royal Astronomical Society*, 425, 3119-3136 (2012), Fig 4 /reuse license acquired from Oxford University Press/
Figure 5.1	[197]	*Monthly Notices of the Royal Astronomical Society Letters*, 407:L6-L10 (2010), Fig 1 /reuse license acquired from Oxford University Press/
Figure 5.2	[198]	*AIP Conference Proceedings*, 822, 75 (2006), Fig 2 /reuse license acquired from AIP Publishing/
Figure 5.3	[198]	*AIP Conference Proceedings*, 822, 75 (2006), Fig 5 /reuse license acquired from AIP Publishing/
Figure 5.4	[198]	*AIP Conference Proceedings*, 822, 75 (2006), Fig 11 /reuse license acquired from AIP Publishing/
Figure 5.5	[204]	*Monthly Notices of the Royal Astronomical Society*, 427, 3081-3102 (2012), Fig 16 /reuse license acquired from Oxford University Press/
Figure 5.6	[207]	*Astrophysical Journal*, 825:4-20, 2016 July 1, Fig 4 /reproduced with permission of Hyewon Suh/

Figure		Source / Credit
Figure 5.7	[215]	*Bulletin of the Astronomical Society of India*, 41:61-115 (2013), Fig 11 /reproduced with permission of D. J. Saikia, BASI/
Figure 5.8	[215]	*Bulletin of the Astronomical Society of India*, 41:61-115 (2013), Fig 12 /reproduced with permission of D. J. Saikia, BASI/
Figure 5.9	[216]	*Monthly Notices of the Royal Astronomical Society*, 447, 2085-2111 (2015), Fig 6 /reuse license acquired from Oxford University Press/
Figure 5.10	[221]	*Science*, 7 January 2005, Volume 307, Issue 5706, pp 77-80, Fig 2 /reuse license acquired from American Association for the Advancement of Science/
Figure 5.11	[232]	*Monthly Notices of the Royal Astronomical Society*, 438, 672-679 (2014), Fig 3 /reuse license acquired from Oxford University Press/
Figure 5.12	[233]	*Astrophysical Journal Letters*, 789:L9-L13, 2014 July 1, Fig 1 /reproduced with permission of Benny Trakhtenbrot/
Figure 5.13		/credit: NASA/CXC/M.Weiss/Natl Astronomical Observatory of Japan/
Figure 5.14	[238]	*Monthly Notices of the Royal Astronomical Society*, 342, 1325-1333 (2003), Figures 1, 4 & 3 /reuse license acquired from Oxford University Press/
Figure 5.15	[239]	*Chinese Journal of Astronomy and Astrophysics*, Volume 5 Supp., 347 (2005), Fig 4 /reproduced with permission from Chinese Journal of Astronomy and Astrophysics/
Figure 5.16	[248]	*Astrophysical Journal*, 536:L5-L9, 2000 June 10, Fig. 2 /reproduced with permission of Jack Sulentic/
Figure 5.17	[249]	*New Astronomy Reviews*, 53:140-148, 2009, Fig 9 /reuse license acquired from Elsevier Science and Technology Journals/
Figure 5.18	[252]	*Astrophysical Journal* 560:L5-L8, 2001 October 10, Fig 2 /reproduced with permission of S. George Djorgovski/
Figure 5.19	[253]	*Astronomical Journal*, 122:2850-2857, 2001 December, Fig 1 /reproduced with permission of Xiaohui Fan/
Figure 5.20	[261]	*Astrophysical Journal*, 722:1312-1324, 2010 October 20, Fig 4 /reproduced with permission of J. Michael Shull/
Figure 5.21	[261]	*Astrophysical Journal*, 722:1312-1324, 2010 October 20, Fig 3 /reproduced with permission of J. Michael Shull/
Figure 6.1	[143]	*Astrophysical Journal Supplement Series*, 80:109-135, 1992 May, Fig 2 /reproduced with permission of Todd Boroson/
Figure 6.2	[143]	*Astrophysical Journal Supplement Series*, 80:109-135, 1992 May, Fig 9 /reproduced with permission of Todd Boroson/
Figure 6.3	[143]	*Astrophysical Journal Supplement Series*, 80:109-135, 1992 May, Fig 5 /reproduced with permission of Todd Boroson/
Figure 6.4	[143]	*Astrophysical Journal Supplement Series*, 80:109-135, 1992 May, Fig 6 /reproduced with permission of Todd Boroson/
Figure 6.5	[270]	*Monthly Notices of the Royal Astronomical Society*, 353, L45-L49 (2004), Fig 2 /reuse license acquired from Oxford University Press/
Figure 6.6	[275]	*Astrophysical Journal*, 733:60-73, 2011 May 20, Fig 9 /reproduced with permission of Jonathan R. Trump/

Figure		Source / Credit
Figure 6.7	[248]	*Astrophysical Journal*, 536:L5-L9, 2000 June 10, Fig 1 /reproduced with permission of Jack Sulentic/
Figure 6.8	[279]	*Annual Review of Astronomy & Astrophysics*, 38:521-571 (2000), Fig 3 /reuse license acquired from EDP Sciences/
Figure 6.9	[285]	*Open Astronomy*, 20:427-434 (2011), Fig 1 /reproduced with permission of Jack Sulentic/
Figure 6.10	[286]	*Astrophysical Journal*, 565:78-85, 2002 January 20, Fig 4 /reproduced with permission of Todd Boroson/
Figure 6.11	[286]	*Astrophysical Journal*, 565:78-85, 2002 January 20, Fig 7 /reproduced with permission of Todd Boroson/
Figure 6.12	[291]	*Monthly Notices of the Royal Astronomical Society*, 403, 1759-1786 (2010), Fig 1*a* /reuse license acquired from Oxford University Press/
Figure 6.13	[291]	*Monthly Notices of the Royal Astronomical Society*, 403, 1759-1786 (2010), Fig 3*b* /reuse license acquired from Oxford University Press/
Figure 6.14	[291]	*Monthly Notices of the Royal Astronomical Society*, 403, 1759-1786 (2010), Fig 4*b* /reuse license acquired from Oxford University Press/
Figure 6.15	[302]	*Nature*, 513, 210-213, 2014 September 11, Fig 1 /reuse license acquired from Springer Nature/
Figure 6.16	[302]	*Nature*, 513, 210-213, 2014 September 11, Fig 2 /reuse license acquired from Springer Nature/
Figure 6.17	[307]	*Astronomy & Astrophysics*, 608, A122 (2017), Fig 1 /© EDP Sciences/
Figure 6.18	[307]	*Astronomy & Astrophysics*, 608, A122 (2017), Fig 9 /© EDP Sciences/
Figure 6.19	[311]	*Monthly Notices of the Royal Astronomical Society*, 443, 1999-2017 (2014), Fig 15 /reuse license acquired from Oxford University Press/
Figure 6.20	[314]	*Astronomical Journal*, 130:367-380, 2005 August, Fig 3 /reproduced with permission of Xiaohui Fan/
Figure 6.21	[315]	*Astrophysical Journal*, 656:680-690, 2007 February 20, Fig 3 /reproduced with permission of Xiaohui Fan/
Figure 6.22		/produced by Robert Gnaski Burton/

Tables

Table		Source / Credit
Table 2.1	[11]	*Astrophysical Journal*, 223:747-757, 1978 August 1, Table 2 /reproduced with permission of Alan Stockton/
Table 2.2	[33]	*Astronomy & Astrophysics*, 243:344-366 (1991), Table 5 /reproduced from data of J. Bergeron and P. Boisse/
Table 2.3	[37]	*Astrophysical Journal*, 566:51-67, 2002 February 10, Table 1 /reproduced with permission of Matthew Malkan/
Table 3.1	[51]	*Astrophysical Journal*, 565:63-77, 2002 January 20, Table 5 /reproduced with permission of Toshinori Maihara/
Table 3.2	[60]	*Astrophysical Journal*, 279:19-39, 1984 April 1, Table 1 /Philipp Kronberg work reproduced with permission of Gary Welter/
Hist. 3.1	[74]	*Astronomical Journal*, 138:1874-1892, 2009 December, Figure 7 /reproduced with permission of M.L. Lister/
Table 3.3	[104]	*Astrophysical Journal*, 648:140-147, 2006 September 1, Table 1 /reproduced with permission of Morley Bell/
Hist. 3.2	[100]	*Astronomical Journal*, 121:21-30, 2001 January, Figure 1*a* /reproduced with permission of W. M. Napier/
Table 3.4	[16]	*Astrophysical Journal*, 706:657-664, 2009 Nov 3, Table 2 /reproduced with permission of W. M. Napier/
Hist. 4.1	[144]	*Astrophysical Journal*, 447:496-504, 1995 July 10, Figure 4 /reproduced from data of Michael Corbin/
Hist. 4.2	[144]	*Astrophysical Journal*, 447:496-504, 1995 July 10, Figure 4 /reproduced from data of Michael Corbin/
Table 4.1	[145]	*Astronomy & Astrophysics*, 407, 461-472 (2003), Table 1 /reproduced from data of W. Kollatschny/
Table 4.2	[211]	*The Physics and Evolution of Active Galactic Nuclei* (Cambridge Univ. Press, 2013), Table 3.1 /reuse license acquired from Cambridge University Press/
Table 4.3	[211]	*The Physics and Evolution of Active Galactic Nuclei* (Cambridge Univ. Press, 2013), Table 3.1 /reuse license acquired from Cambridge University Press/
Table 4.4		/produced by Robert Gnaski Burton/
Table 4.5		/produced by Robert Gnaski Burton/
Table 5.1	[212]	*Monthly Notices of the Royal Astronomical Society*, 373, 613-622 (2006), Table 1 /reuse license acquired from Oxford University Press/
Table 5.2	[245]	*Physica Scripta*, Vol. 29, 543-550, 1984, Table 2 /reuse license acquired from IOP Publishing/
Table 5.3	[245]	*Physica Scripta*, Vol. 29, 543-550, 1984, Table 3 /reuse license acquired from IOP Publishing/
Table 6.1	[143]	*Astrophysical Journal Supplement Series*, 80:109-135 (1992), Table 4 /reproduced with permission of Todd Boroson/
Table 6.2	[265]	*Astrophysical Journal*, 403:L9-L11, 1993 January 20, Table 1 /reproduced from data of Michael Corbin/
Table 6.3	[267]	*Astrophysical Journal Supplement Series*, 107:69-96 (1996), Table 8 /reproduced with permission of Todd Boroson/

Appendix B – Chronology of 4DE1

Jack Sulentic, Paola Marziani, and collaborators produced a wealth of information on active galactic nuclei and the 4DE1 parameter space. The following list highlights three decades of their work. Reference numbers [1] through [315] appear in the text. See **References** for title and journal information.

- (Sulentic, 1989) The broad-line Hβ profile displacements show evidence for a small gravitational redshift in most AGNs. [318]
- (Marziani *et al.*, 1993) The ratio of intensities between the broad and narrow Hβ components in radio galaxy OQ 208 increased monotonically from ~15 to ~40 over a period of six years. A correlation exists between the Hβ profile broad peak and its luminosity. [319]
- (Marziani *et al.*, 1996) RL objects usually show redshifted asymmetric Hβ and unshifted symmetric $C_{IV}\lambda1549$; RQ objects usually show blueshifted asymmetric C_{IV} and unshifted symmetric Hβ. [320]
- (Sulentic *et al.*, 1998) Evidence for and against an accretion disk origin for the Fe Kα emission is equivocal. [321]
- (Sulentic & Marziani, 1999) BLR properties inferred from C_{IV} profiles require subtraction of the narrow component. [322]
- (Sulentic *et al.*, 2000a) Correlation between FWHM (Hβ), $R(Fe_{II})$ and the soft X-ray spectral index provides optimal discrimination between all principal AGN types (from NLS1 to radio galaxies). [279]
- (Sulentic *et al.*, 2000b) Observations of PG 1416–129 confirm the existence of a redshifted VBLR component distinct from the broad component. [323]
- (Sulentic *et al.*, 2000c) An ideal three-dimensional eigenvector (3DE1) correlation space for AGNs consists of FWHM (Hβ), R Fe_{II}, and the soft X-ray photon index Γ_{soft}. The C_{IV} blueshifts are strongest in AGNs with high L/L_{Edd}. [248]
- (Marziani *et al.*, 2001) The physical driver for the 4DE1 space is L/L_{Edd}. An anticorrelation exists between L/L_{Edd} and FWHM(Hβ_{BC}). [278]
- (Sulentic *et al.*, 2002) Hβ broad-line profiles (FWHM, centroid shift, and profile asymmetry) change along the E1 sequence. Objects with FWHM(Hβ_{BC}) > 4000 km s^{-1} are best fit with two components, BC and redshifted VBLR. [281]
- (Zamanov & Marziani, 2002) Accreting white dwarfs (nanoquasars?) show strong spectral similarities to quasars. [324]
- (Krongold *et al.*, 2002) An evolutionary sequence from type-2 to type-1 AGN may be appropriate for most AGN and may be independent on luminosity. [289]
- (Zamanov *et al.*, 2002) Direct kinematic linkage exists between narrow- and broad-line-emitting gases. The blueshifting of [O_{III}] and C_{IV} are correlated. [282]

- (Marziani *et al.*, 2003a) Hβ region features for 215 type-1 AGNs (z < 0.8) show that the HIL and LIL emit from two distinct emitting regions. [280]
- (Marziani *et al.*, 2003b) RQ sources have systematically smaller masses and larger L/L_{Edd}. Source orientation plays an increasingly important role as FWHM(Hβ_{BC}) increases. [287]
- (Sulentic *et al.*, 2003) The RL/RQ dichotomy may be more related to evolution than to fundamentally different BLR structure. [325]
- (Bachev *et al.*, 2004) BH spin, rather than BH mass or accretion rate, is a key trigger in determining whether an object will be RL or RQ. Models involving (blue-shifted) line-emitting clouds suffer a number of difficulties. [303]
- (Sulentic *et al.*, 2004) The E1 optical parameters show no dependence on luminosity to at least $z < 1$. A systematic increase in *minimum* Hβ width with luminosity occurs for intermediate redshift ($0.85 < z < 2.5$) sources. [298]
- (Sulentic *et al.*, 2006) Line asymmetries and redshifted VBLR components introduce errors into M_{BH} derivations. It is not certain that FWHM(Hβ_{BC}) is a valid estimator of the virial velocity even after proper subtraction of Hβ_{NC}, Fe_{II} and $[O_{III}]\lambda\lambda4959,5007$. [146]
- (Sulentic *et al.*, 2007) Systematic blueshift of $C_{IV}\lambda1549$ is physically orthogonal to previously defined E1 parameters. This fourth element of the eigenvector space (4DE1) is not observed in broader line AGNs (including most RL sources). [283]
- (Sulentic *et al.*, 2008) Diagnostic power of 4DE1 is confirmed using optical spectra from the SDSS, UV spectra from HST, and X-ray spectra from XMM. [326]
- (Zamfir *et al.*, 2008) 4DE1 parameters show no obvious dependence on source luminosity. They also show no dependence on radio luminosity apart from a restricted domain space occupation observed only for RL sources. [288]
- (Marziani *et al.*, 2009) Intermediate redshift ($0.9 < z < 3.0$) sources show: (1) higher values of FWHM(Hβ_{BC}) and R_{FeII}; (2) decrease in equivalent widths of Hβ_{BC} and $[O_{III}]\lambda\lambda4959,5007$ with increasing source bolometric luminosity; (3) increase in Hβ redward asymmetry. [300]
- (Sulentic *et al.*, 2009) Inferred M_{BH} estimates are not large enough to make it likely that the VBLR arises close to the black hole. [327]
- (Zamfir *et al.*, 2010) Line profiles of low-*z* Hβ emission lines show large differences along the AGN main sequence. A refinement of the 4DE1 space can be provided by separating the two populations of quasars at $R_{FeII} \sim 0.50$ rather than at FWHM(Hβ) = 4000 km s^{-1}. [291]
- (Marziani *et al.*, 2010) Population A quasars are dominated by BC emission in Hβ and BLUE component emission in $C_{IV}\lambda1549$. The most distinguishing feature of population B quasars is the VBC with high ionization and large column density. FWHM($Mg_{II}\lambda2800$) ≥ FWHM(Hβ) in the highest Eddington radiators, and it shows a significant blueshift relative to Hβ. [328]
- (Dultzin *et al.*, 2011) Properties seem preserved for NLS1 sources at high redshift. [296]
- (Sulentic *et al.*, 2011) Differences in population A/B are more fundamental than the RL/RQ dichotomy. [285]

- (Sulentic *et al.*, 2012) No evidence exists for a systematic redshift of Fe_{II}. [329]
- (Negrete *et al.*, 2012) Inferred conditions for BLR include a low ionization parameter (less than 10^{-2}), high density (10^{12}-10^{13} cm^{-3}) and significant metal enrichment. The product of density and ionization parameter better estimates the BLR radius, hence M_{BH}. [297]
- (Marziani & Sulentic, 2012) HIL blueshifts are observed in both the BLR and NLR, while the LIL appears more symmetric. C_{IV} blueshifts and maximum amplitude decrease along the E1 sequence toward Pop. B. The $C_{IV}\lambda1549$ and $Mg_{II}\lambda2800$ blueshifts are likely related. [304]
- (Marziani *et al.*, 2013) FWHM($Mg_{II}\lambda2800$) is equal to or greater than FWHM(Hβ) in the highest Eddington radiators and shows a significant blueshift relative to Hβ. [330]
- (Marziani & Sulentic, 2014a) Sources radiating at L/L_{Edd} ~ 1 may be prime candidates as cosmological probes capable of yielding meaningful constraints on Ω_M and Ω_Λ. [331]
- (Marziani *et al.*, 2014) Population A/B differences could be associated with an accretion mode transition, with high L/L_{Edd} sources accreting in an advection-dominated flow. [310]
- (Marziani & Sulentic, 2014b) Neither luminosity nor mass is a main driver of quasar spectral diversity. [294]
- (Sulentic *et al.*, 2014a) After fifty years, 4DE1 remains the best basis for quasar models. [332]
- (Sulentic *et al.*, 2014b) GTC high-*z* objects show the same *relative* differences in $C_{IV}\lambda1549$ profiles between population A and B sources. High-*z* objects show the same trends for M_{BH} and L/L_{Edd} found in low-*z* objects. [306]
- (Marziani *et al.*, 2015) UV spectral analysis renders possible the density and ionization parameter at the two extremes of the 4DE1 sequence. [292]
- (Sulentic & Marziani, 2015) The historical development of 4DE1 points toward an increasing ability to systematically organize quasar properties. [333]
- (Bensch *et al.*, 2015) Significant differences appear in soft X-ray excess between pop-A and pop-B (z < 0.8) objects. [293]
- (Marziani *et al.*, 2016a) High-velocity outflows are more frequent as redshift increases. [305]
- (Marziani *et al.*, 2016b) [O_{III}]λ5007 'blue outliers' are much more common at intermediate-*z*. [301]
- (Sulentic *et al.*, 2017) C_{IV} outflows become stronger and more frequent in Hamburg ESO (HE) high-*z*, high luminosity sources. [307]
- (Fraix-Burnet *et al.*, 2017) The phylogeny of low-*z* quasars and the ontogeny of their central black holes clarify the evolution of the 4DE1 parameter space. [295]
- (Marziani *et al.*, 2018) Best evidence shows that pop-A and pop-B represent different structures. The relative balance between gravitational and radiation forces (*i.e.*, L/L_{Edd}) appears as a major factor in both the dynamics and the physical conditions of the line emitting gas. [334]

References

Chapter 1

[1] Hubble, Edwin P., *The Realm of the Nebulae* (Yale University Press, 1936).

[2] Bothun, G., *Modern Cosmological Observations and Problems* (Taylor & Francis Ltd., 1998).

[3] Freedman, W. and Madore, B.F, "The Hubble Constant." *Annual Review of Astronomy & Astrophysics,* Volume 48, 2010.

[4] Riess *et al.*, "Observational Evidence from Supernovae for an Accelerating Universe and a Cosmological Constant." *Astronomical Journal,* 116:1009-1038, 1998 September.

[5] Perlmutter *et al.*, "Measurements of Omega and Lambda from 42 High-Redshift Supernovae." *Astrophysical Journal,* 517:565-586, 1999 June.

[6] Kragh, H.S. and Overduin, J.M., *The Weight of the Vacuum: A Scientific History of Dark Energy* (Springer, 2014).

[7] Kurki-Suonio, H., "Lecture Notes for Cosmology I + II." <http://www.helsinki.fi/~hkurkisu/cosmology/Cosmo4.pdf> (2009).

Chapter 2

[8] *Report of the Research Briefing Panel on Astronomy and Astrophysics* (National Academy Press, 1983).

[9] Hoyle, Burbidge and Narlikar, *A Different Approach to Cosmology* (Cambridge University Press, 2000).

[10] Kellermann, K.I., "The Discovery of Quasars and its Aftermath." *Journal of Astronomical History and Heritage,* 17(3), 267-282 (2014).

[11] Stockton, A., "The Nature of QSO Redshifts." *Astrophysical Journal,* 223:747-757, 1978 August 1.

[12] Peterson, Bradley M., *Active Galactic Nuclei* (Cambridge University Press, 1997), p. 133.

[13] Yee, H.K.C. and Green, R.F., "An Imaging Survey of Fields Around Quasars. II. The Association of Galaxies with Quasars." *Astrophysical Journal,* 280:79-90, 1984 May 1.

[14] Gehren *et al.*, "Host Galaxies of Quasars and Their Association with Galaxy Clusters." *Astrophysical Journal,* 278:11-27, 1984 March 1.

[15] Hutchings *et al.*, "Optical Imaging of 78 Quasars and Host Galaxies." *Astrophysical Journal,* 280:41-50, 1984 May 1.

[16] Burbidge G. and Napier, W.M., "Associations of High-Redshift Quasi-Stellar Objects with Active, Low-Redshift Spiral Galaxies." *Astrophysical Journal,* 706:657-664, 2009 Nov 20.

[17] Gorenstein *et al.*, "The Milli-Arcsecond Images of Q0957+561." *Astrophysical Journal,* 287:538-548, 1984 December 15.

[18] van der Wel *et al.*, "Discovery of a Quadruple Lens in CANDELS with a Record Lens Redshift $z = 1.53$." *Astrophysical Journal,* 777:L17-L21, 2013 November 1.

[19] Wong *et al.*, "Discovery of a Strong Lensing Galaxy Embedded in a Cluster at $z = 1.62$." *Astrophysical Journal,* 789:L31-L36, 2014 July 10.

[20] Kochanek, C.S., Falco, E.E., Impey, C., Lehar, J., Mcleod, B., Rix, H.-W., "Gravitational Lens Database." <https://www.cfa.harvard.edu/castles/noimages.html>.

[21] Sargent *et al.*, "The Distribution of Lyman-Alpha Absorption Lines in the Spectra of Six QSOs: Evidence for an Intergalactic Origin." *Astrophysical Journal Supplement Series,* 42:41-81, 1980 January.

[22] Young *et al.*, "C IV Absorption in an Unbiased Sample of 33 QSOs: Evidence for the Intervening Galaxy Hypothesis." *Astrophysical Journal Supplement Series,* 48:455-506, 1982 April.

[23] Weymann *et al.*, "Absorption Lines in the Spectra of Quasi-Stellar Objects." *Annual Review of Astronomy & Astrophysics,* 19, 41-76 (1981).

[24] Haschick, A.D. and Burke, B.F., "Neutral Hydrogen Absorption in the Spectrum of the Quasar-Galaxy Pair 4C 32.33/NGC 3067." *Astrophysical Journal*, 200:L137-L140, 1975 September 15.

[25] Peterson, B.M., "Neutral Hydrogen Absorption in the Spectra of Quasi-Stellar Objects: A Search for Absorption due to Clusters of Galaxies." *Astrophysical Journal*, 223:740-746, 1978 August 1.

[26] Richards *et al.*, "Determining the Fraction of Intrinsic C IV Absorption in Quasi-Stellar Object Absorption-Line Systems." *Astrophysical Journal*, 513:576-591, 1999 March 10.

[27] Libby *et al.*, "Systematics of Quasi-Stellar-Object Spectra." *Astronomical Journal*, 89:311-315, 1984 March.

[28] Roberts *et al.*, "A Study of the Absorption Line Spectra of Six High-Redshift Quasi-Stellar Objects." *Astrophysical Journal*, 224:344-367, 1978 September 1.

[29] Young *et al.*, "A High-Resolution Study of the Absorption Spectrum of PKS 2126-158." *Astrophysical Journal*, 229:891-908, 1979 May 1.

[30] Sargent *et al.*, "A High-Resolution Study of the Absorption Spectra of the QSOs Q0002-422 and Q0453-423." *Astrophysical Journal*, 230:49-67, 1979 May 15.

[31] Sargent *et al.*, "A High-Resolution Spectroscopic Study of Q0119-046 and the Nature of Absorption Complexes with $z_{abs} > z_{em}$." *Astrophysical Journal*, 252:54-68, 1982 January 1.

[32] Varshni, Y.P. and Singh, D., "Common Lines in the Rest-Frame Absorption-Line Spectra of QSOs?" *Astrophysics & Space Science*, Volume 109, Issue 1, pp 149-153, February 1985.

[33] Bergeron, J. and Boisse, P., "A Sample of Galaxies Giving Rise to Mg II Quasar Absorption Systems." *Astronomy and Astrophysics*, 243, 344-366 (1991).

[34] LeBrun *et al.*, "A Deep Imaging Survey of Fields Around Quasars with $z < 1.2$ Mg II Absorption Systems." *Astronomy & Astrophysics*, 279, 33-52 (1993).

[35] Morris *et al.*, "First Results from the Goddard High-Resolution Spectrograph: The Galactic Halo and the Lyα Forest at Low Redshift in 3C 273." *Astrophysical Journal*, 377:L21-L24, 1991 August 10.

[36] Lanzetta *et al.*, "The Gaseous Extent of Galaxies and the Origin of Lyman-Alpha Absorption Systems: A Survey of Galaxies in the Fields of *Hubble Space Telescope* Spectroscopic Target QSOs." *Astrophysical Journal*, 442:538-568, 1995 April 1.

[37] Colbert, J.W. and Malkan, M.A., "NICMOS Snapshot Survey of Damped Lyα Quasars." *Astrophysical Journal*, 566:51-67, 2002 February 10.

Chapter 3

[38] Kembhavi and Narlikar, *Quasars and Active Galactic Nuclei* (Cambridge University Press, 1999), Fig. 7.3.

[39] Farrah *et al.*, "The X-Ray Spectrum of the $z = 6.30$ QSO SDSS J1030+0524." *Astrophysical Journal*, 611:L13-L16, 2004 August 10.

[40] Fan, X., "Evolution of High-Redshift Quasars." *New Astronomy Reviews*, 50:665-671, 2006.

[41] Djorgovski, *et al.*, "The Origins and the Early Evolution of Quasars and Supermassive Black Holes." <http://arxiv.org/abs/0803.2862>, arXiv:0803.2862v1 [astro-ph], 19 Mar 2008.

[42] Wu *et al.*, "An Ultraluminous Quasar with a Twelve Billion Solar Mass Black Hole at Redshift 6.30." *Nature*, 518:512-515, 26 February 2015.

[43] Kundt, W., "Critical Thoughts on Cosmology." *AIP Conference Proceedings*, 1132, 288 (2009).

[44] Hawking, S., "Black Hole Explosions?" *Nature*, 248:30-31, 1 March 1974.

[45] Dietrich *et al.*, "Quasar Elemental Abundances at High Redshifts." *Astrophysical Journal*, 589:722-732, 2003 June 1.

[46] Fan *et al.*, "A Survey of $z > 5.8$ Quasars in the Sloan Digital Sky Survey. I. Discovery of Three New Quasars and the Spatial Density of Luminous Quasars at $z \sim 6$." *Astronomical Journal*, 122:2833-2849, 2001 December.

[47] Constantin *et al.*, "Emission-Line Properties of $z > 4$ Quasars." *Astrophysical Journal*, 565:50-62, 2002 January 20.

[48] Thompson *et al.*, "Lack of Iron Abundance Evolution in High Redshift QSOs." *Astrophysical Journal*, 515:487-496, 1999 April 20.

[49] Maiolino *et al.*, "Early Star Formation Traced by the Highest Redshift Quasars." *Astrophysical Journal*, 596:L155-L158, 2003 October 20.

[50] Barth *et al.*, "Iron Emission in the $z = 6.4$ Quasar SDSS J114816.64+525150.3." *Astrophysical Journal*, 594:L95-L98, 2003 September 10.

[51] Iwamuro *et al.*, "FE $_{II}$/Mg $_{II}$ Emission-Line Ratios of QSOs Within $0 < z < 5.3$." *Astrophysical Journal*, 565:63-77, 2002 January 20.

[52] Freudling *et al.*, "Iron Emission in $z \sim 6$ QSOs." *Astrophysical Journal*, 587:L67-L70, 2003 April 20.

[53] Reeves, J.N. and Turner, M.J.L., "X-Ray Spectra of a Large Sample of Quasars with ASCA." *Monthly Notices of the Royal Astronomical Society*, 316, 234 (2000).

[54] Vignali *et al.*, "X-Ray Lighthouses of the High-Redshift Universe. II. Further Snapshot Observations of the Most Luminous $z > 4$ Quasars with Chandra." *Astronomical Journal*, 129:2519-2530, 2005 June.

[55] Maiolino, R. and Risaliti, G., "X-ray Absorption in Active Galactic Nuclei." *ASP Conference Series*, Volume 373 (The Central Engine of AGN), 2007, L. Ho and J. Wang, (editors).

[56] Archibald *et al.*, "A Submillimetre Survey of the Star Formation History of Radio Galaxies." *Monthly Notices of the Royal Astronomical Society*, 323, 417-444 (2001).

[57] Bertoldi *et al.*, "Dust Emission from the Most Distant Quasars." *Journal of Astronomy and Astrophysics*, 406, L55-L58 (2003).

[58] Dunne *et al.*, "Type II Supernovae as a Significant Source of Interstellar Dust." *Nature*, 424:285-287, 17 July 2003.

[59] Rowlands *et al.*, "The Dust Budget Crisis in High-Redshift Submillimetre Galaxies." *Monthly Notices of the Royal Astronomical Society*, 441, 1040-1058 (2014).

[60] Welter, G.L., Perry, J.J. and Kronberg, P.P., "The Rotation Measure Distribution of QSOs and of Intervening Clouds: Magnetic Fields and Column Densities." *Astrophysical Journal*, 279:19-39, 1984 April 1.

[61] Arp, H., "Intergalactic Magnetic Field and the Distance of Quasars from Faraday Rotation." *Physics Letters A*, Volume 129, Issue 3, pp 135-139, 16 May 1988.

[62] Goldhaber *et al.*, "Timescale Stretch Parameterization of Type Ia Supernova *B*-Band Light Curves." *Astrophysical Journal*, 558:359-368, 2001 September 1.

[63] Hawkins, M.R.S., "On Time Dilation in Quasar Light Curves." *Monthly Notices of the Royal Astronomical Society*, 405, 1940-1946 (2010).

[64] Rourke, C., "Intrinsic Redshift in Quasars." <citeseerx.ist.psu.edu/viewdoc/download?doi=10.1.1.697.9935&rep=rep1&type=pdf>.

[65] <http://www.astro.caltech.edu/~george/ay127/readings/KraussChaboyer2003.pdf>.

[66] Komossa, S. and Hasinger, G., "The X-Ray Evolving Universe: (Ionized) Absorption and Dust, from Nearby Seyfert Galaxies to High-Redshift Quasars." <arxiv.org/pdf/astro-ph/0207321v1.pdf>, arXiv-ph:astro-ph/0207321v1, 15 July 2002.

[67] Hamann, F. and Ferland, G., "The Chemical Evolution of QSOs and the Implications for Cosmology and Galaxy Formation." *Astrophysical Journal*, 418:11-27, 1993 November 20.

[68] Friaca *et al.*, "An Old Quasar in a Young Dark Energy-Dominated Universe?" Monthly *Notices of the Royal Astronomical Society*, 362, 1295-1300 (2005).

[69] Jain, D. and Dev, A., "Age of High Redshift Objects – A Litmus Test for the Dark Energy Models." *Physics Letters B*, 633, pp 436-440 (2006).

[70] Clark, S., *Redshift* (University of Hertfordshire Press, 1997), Chapter 5.

[71] Narlikar, J.V. and Chitre, S.M., "Faster-Than-Light Motion in Quasars." *Journal of Astrophysics and Astronomy*, 5:495-506 (1984).

[72] Procas, R., "Superluminal Motions: Astronomers Still Puzzled." *Nature*, 302:753-754, 28 April 1983.

[73] Thomson, R.C., Mackay, C.D., and Wright, A.E., "Internal Structure and Polarization of the Optical Jet of the Quasar 3C273." *Nature*, 365:133-135, 9 September 1993.

[74] Lister *et al.*, "MOJAVE: Monitoring of Jets in Active Galactic Nuclei with VLBA Experiments. VI. Kinematics Analysis of a Complete Sample of Blazar Jets." *Astronomical Journal*, 138:1874-1892, 2009 December.

[75] Meyer *et al.*, "An *HST* Proper-Motion Study of the Large-Scale Jet of 3C273." *Astrophysical Journal*, Volume 818, 195, 2016 February 18.

[76] Lister *et al.*, "MOJAVE. X. Parsec-Scale Jet Orientation Variations and Superluminal Motion in Active Galactic Nuclei." *Astronomical Journal*, 146:120, 2013 November.

[77] Tifft, W.G., "Properties of the Redshift-Magnitude Bands in the Coma Cluster." *Astrophysical Journal*, 179:29-44, 1973 January 1.
[78] Tifft, W.G., "Discrete States of Redshift and Galaxy Dynamics. I. Internal Motions in Single Galaxies." *Astrophysical Journal*, 206:38-56, 1976 May 15.
[79] Tifft, W.G., "Discrete States of Redshift and Galaxy Dynamics. II. Systems Of Galaxies." *Astrophysical Journal*, 211:31-46, 1977 January 1.
[80] Tifft, W.G., "Discrete States of Redshift and Galaxy Dynamics. III. Abnormal Galaxies and Stars." *Astrophysical Journal*, 211:377-391, 1977 January 15.
[81] Tifft, W.G., "Periodicity in the Redshift Intervals for Double Galaxies." *Astrophysical Journal*, 236:70-74, 1980 February 15.
[82] Tifft, W.G., "Double Galaxy Investigations. I. Observations." *Astrophysical Journal Supplement Series*, 50:319-390, 1982 November.
[83] Tifft, W.G., " Quantum Effects in the Redshift Intervals for Double Galaxies." *Astrophysical Journal*, 257:442-449, 1982 June 15.
[84] Tifft, W.G., "Double Galaxy Investigations. II. The Redshift Periodicity in Optically Observed Pairs." *Astrophysical Journal*, 262:44-47, 1982 November 1.
[85] Tifft, W.G., "Double Galaxy Investigations. III. The Differential Redshift Distribution and Emission-Line Correlations." *Astrophysical Journal*, 288:65-72, 1985 January 1.
[86] Tifft, W.G. and Cocke, W.J., "Global Redshift Quantization." *Astrophysical Journal*, 287:492-502, 1984 December 15.
[87] Croasdale, M.R., "Periodicities in Galaxy Redshifts." *Astrophysical Journal*, 345:72-83, 1989 October 1.
[88] Guthrie, B.N.G. and Napier, W.M., "The Virgo Cluster as a Test for Quantization of Extragalactic Redshifts." *Monthly Notices of the Royal Astronomical Society*, 243, 431-442, April 1, 1990.
[89] Guthrie, B.N.G. and Napier, W.M., "Evidence for Redshift Periodicity in Nearby Field Galaxies." *Monthly Notices of the Royal Astronomical Society*, 253, 533-544 (1991).
[90] Guthrie, B.N.G. and Napier, W.M., "Redshift Periodicity in the Local Supercluster." *Astronomy & Astrophysics*, 310, 353-370 (1996).
[91] Koo *et al.*, "Deep Pencil-Beam Redshift Surveys as Probes of Large Scale Structures." *ASP Conference Series*, Volume 51 (Observational Cosmology), 1993, G. Chincarini, A. Iovino, T. Maccacaro, and D. Maccagni (editors).
[92] Broadhurst *et al.*, "Large-Scale Distribution of Galaxies at the Galactic Poles." *Nature*, 343:726-728, 22 February 1990.
[93] Saar *et al.*, "The Supercluster-Void Network V. The Regularity Periodogram." *Astronomy & Astrophysics*, 393, 1-23 (2002).
[94] Burbidge, G. and Burbidge, M., *Quasi-Stellar Objects* (W.H. Freeman & Co., 1967).
[95] Karlsson, K.G., "On the Existence of Significant Peaks in the Quasar Redshift Distribution." *Astronomy & Astrophysics*, 58, 237-240 (1977).
[96] Fang *et al.*, "The Periodicity in the Distribution of Quasar Redshifts and the Density Perturbations in the Early Universe." *Astronomy & Astrophysics*, 106, 287-292 (1982).
[97] Duari, *et al.*, "Statistical Tests of Peaks and Periodicities in the Observed Redshift Distribution of Quasi-Stellar Objects." *Astrophysical Journal*, 384:35-42, 1992 January 1.
[98] Burbidge, G.R., "The Distribution of Redshifts in Quasi-Stellar Objects, N-Systems, and Some Radio and Compact Galaxies." *Astrophysical Journal*, 154:L41-L48, November 1968.
[99] Bell, M.B., "Distances of Quasars and Quasar-Like Galaxies: Further Evidence that Quasi-Stellar Objects may be Ejected from Active Galaxies." *Astrophysical Journal*, 616:738-744, 2004 December 1.
[100] Burbidge, G. and Napier, W.M., "The Distribution of Redshifts in New Samples of Quasi-Stellar Objects." *Astronomical Journal*, 121:21-30, 2001 January.
[101] Hawkins *et al.*, "No Periodicities in 2dF Redshift Survey Data." *Monthly Notices of the Royal Astronomical Society*, 336, L13-L16 (2002).
[102] Napier, W.M. and Burbidge, G., "The Detection of Periodicity in QSO Data Sets." *Monthly Notices of the Royal Astronomical Society*, 342, 601-604 (2003).
[103] Tang, S.M. and Zhang, S.N., "Critical Examinations of QSO Redshift Periodicities and Associations with Galaxies in Sloan Digital Sky Survey Data." *Astrophysical Journal*, 633:41-51, 2005 November 1.

[104] Bell, M.B. and McDiarmid, D., "Six Peaks Visible in the Redshift Distribution of 46,400 SDSS Quasars Agree with the Preferred Redshifts Predicted by the Decreasing Intrinsic Redshift Model." *Astrophysical Journal*, 648:140-147, 2006 September 1.

[105] Repin *et al.*, "Absence of a Periodic Component in the Quasar *z* Distribution." *Astronomy Reports*, Volume 56, Issue 9, pp 702-709, September 2012.

[106] Fulton, C. "Physical Association and Periodicity in Quasar Families with SDSS and SMRS." *Astrophysics & Space Science*, 363:134, July 2018.

[107] Arp, H. and Russell, D., "A Possible Relationship Between Quasars and Clusters of Galaxies." *Astrophysical Journal*, 549:802-819, 2001 March 10.

[108] Arp, H., *Seeing Red: Redshifts, Cosmology and Academic Science* (Apeiron, 1998).

[109] Sulentic, J., "Confirmation of the Luminous Connection Between NGC 4319 and Markarian 205." *Astrophysical Journal*, 265:L49-L53, 1983 February 15.

[110] Lopez-Corredoira, M. and Gutierrez, C.M., "Two Emission Line Objects with $z > 0.2$ in the Optical Filament Apparently Connecting the Seyfert Galaxy NGC 7603 to its Companion." *Astronomy & Astrophysics*, 390, L15-L18 (2002).

[111] Galianni, Burbidge, Arp, Junkkarinen, Burbidge and Zibetti, "The Discovery of a High-Redshift X-Ray-Emitting QSO very Close to the Nucleus of NGC 7319." *Astrophysical Journal*, 620:88-94, 2005 February 10.

[112] Chu *et al.*, "Quasars Around the Seyfert Galaxy NGC 3516." *Astrophysical Journal*, 500:596-598, 1998 June 20.

[113] Weedman, Daniel W., Qu*asar Astronomy* (Cambridge University Press, 1986).

[114] Burbidge, E.M., Burbidge, G.R., Fowler, W.A., and Hoyle, F., "Synthesis of the Elements in Stars." *Reviews of Modern Physics* 29, 547, 1957 October 1.

[115] Burbidge *et al.*, "QSOs and Active Galactic Nuclei Associated with NGC 2639." *Astrophysical Journal Supplement Series*, 153:159-163, 2004 July.

[116] Burbidge *et al.*, "Apparent Associations Between Bright Galaxies and Quasi-Stellar Objects." *Astrophysical Journal*, 170:233-240, 1971 December 1.

[117] Burbidge, G.R., Narlikar, J.V., and Hewitt, A., "The Statistical Significance of Close Pairs of QSOs." *Nature*, 317:413-415, 3 October 1985.

[118] Sulentic, J.W. and Tifft, W.G., *The Revised New General Catalog Of Nonstellar Astronomical Objects* (The University of Arizona Press, 1973).

[119] Burbidge, G., Hewitt, A., Narlikar, J.V., and Das Gupta, P., "Associations Between Quasi-Stellar Objects and Galaxies." *Astrophysical Journal Supplement Series*, 74:675-730, 1990 November.

[120] Burbidge, G., "The Reality of Anomalous Redshifts in the Spectra of Some QSOs and its Implications." *Astronomy & Astrophysics*, 309, 9-22 (1996).

[121] Zhu, X.-F. and Chu, Y.-Q., "The Association Between Quasars and the Galaxies of the Virgo Cluster." *Astronomy & Astrophysics*, 297, 300-304 (1995).

[122] Canizares, C., "Gravitational Focusing and the Association of Distant Quasars with Foreground Galaxies." *Nature*, 291:620-624, 25 June 1981.

Chapter 4

[123] Burbidge G., "A Realistic Cosmological Model Based on Observations and Some Theory Developed over the Last 90 Years." arXiv:0811.2402v1 [astro-ph], 14 Nov 2008.

[124] Ratcliffe H., "A Review of Anomalous Redshift Data." *CC-2 ASP Conference Series*, Vol. 413 (2009), Frank Potter, ed.

[125] Narlikar, J.V. and Arp, H., "Flat Spacetime Cosmology: A Unified Framework for Extragalactic Redshifts." *Astrophysical Journal*, 405:51-56, 1993 March 1.

[126] Narlikar, J.V., "Anomalous Redshifts and the Variable Mass Hypothesis." *Astrophysics & Space Science*, Volume 244, Issue 1, pp 177-186, March 1996.

[127] Tifft, W. G., *Redshift, Key to Cosmology* (W.G. Tifft, 2014).

[128] Russell, D.G., "Evidence for Intrinsic Redshifts in Normal Spiral Galaxies." *Astrophysics & Space Science*, Volume 298, Issue 4, pp 577-602, August 2005.

[129] Greenstein, J. L. and Schmidt, M., "The Quasi-Stellar Radio Sources 3C48 and 3C273." *Astrophysical Journal*, 140:1-34, 1964 July 1.

[130] Bondi, H., "Massive Spheres in General Relativity." *Proceedings of the Royal Society of London*, 282:1390, pp 303-317, 24 November 1964.

[131] Hoyle, F. and Fowler, W.A., "Gravitational Red-Shifts in Quasi-Stellar Objects." *Nature*, 213:373-374, 28 January 1967.

[132] Ipser, J.R., "Relativistic, Spherically Symmetric Star Clusters III. Stability of Compact Isotropic Models." *Astrophysical Journal*, 158:17-43, October 1969.

[133] Fackerell *et al.*, "Relativistic Star Clusters." *Comments on Astrophysics and Space Physics*, Volume 1, page 134 (1969).

[134] Das, P.K., "Physical Properties of Collapsed Objects with Large Central Gravitational Redshifts." *Monthly Notices of the Royal Astronomical Society*, 177, 391-408 (1976).

[135] Yabushita, S., "On Dynamical Stability of Supermassive Objects." *Monthly Notices of the Royal Astronomical Society*, 174, 637-647 (1976).

[136] Rasio *et al.*, "On the Existence of Stable Relativistic Star Clusters with Arbitrarily Large Central Redshifts." *Astrophysical Journal*, 336:L63-L66, 15 January 1989.

[137] Kogan *et al.*, "Stability of Dense Stellar Clusters Against Relativistic Collapse. II. Maxwellian Distribution Functions with Different Cutoff Parameters." *Astrophysical Journal*, 500:217-232, 1998 June 10.

[138] Thorne, K., *Black Holes and Time Warps* (W.W. Norton & Company, 1994).

[139] Silk, J., *On the Shores of the Unknown* (Cambridge University Press, 2005), page 143.

[140] Hoyle, F., "A Gravitational Model for QSOs." *Astrophysics and Relativity Preprint Series 88*, University College, Cardiff (1983).

[141] Abbott *et al.*, "Observation of Gravitational Waves from a Binary Black Hole Merger." *Physical Review Letters*, 116:061102, 12 February 2016.

[142] Popovic *et al.*, "Contribution of Gravitational Redshift to Spectral Line Profiles of Seyfert Galaxies and Quasars." *Astronomy & Astrophysics*, 293, 309-314 (1995).

[143] Boroson, T.A. and Green, R.F., "The Emission-Line Properties of Low-Redshift Quasi-Stellar Objects." *Astrophysical Journal Supplement Series*, 80:109-135, 1992 May.

[144] Corbin, M.R., "QSO Broad Emission Line Asymmetries: Evidence of Gravitational Redshift?" *Astrophysical Journal*, 447:496-504, 1995 July 10.

[145] Kollatschny, W., "Accretion Disk Wind in the AGN Broad-Line Region: Spectroscopically Resolved Line Profile Variations in Mrk 110." *Astronomy & Astrophysics*, 407, 461-472 (2003).

[146] Sulentic *et al.*, "VLT/ISAAC Spectra of the Hβ Region in Intermediate Redshift Quasars II. Black Hole Mass and Eddington Ratio." *Astronomy & Astrophysics*, 456, 929-939 (2006).

[147] Peterson, B.M., "Measuring the Masses of Supermassive Black Holes." *Space Science Reviews*, Volume 183, Issue 1-4, pp 253-275, September 2014.

[148] Bon *et al.*, "Gravitational Redshift of Emission Lines in the AGN Spectra." *Astrophysics & Space Science*, 360:41, December 2015.

[149] Muller, A. and Wold, M., "On The Signatures of Gravitational Redshift: The Onset of Relativistic Emission Lines." *Astronomy & Astrophysics*, 457:485-492 (2006).

[150] Dubey, A.K. and Sen, A.K., "An Analysis of Gravitational Redshift from Rotating Body." *International Journal of Theoretical Physics*, Volume 54, Issue 7, pp 2398-2417, July 2015.

[151] Taylor, J.H. and Weisberg, J.M., "Further Experimental Tests of Relativistic Gravity Using the Binary Pulsar PSR 1913+16." *Astrophysical Journal*, 345:434-450, 1989 October 1.

[152] Volonteri *et al.*, "The Distribution and Cosmic Evolution of Massive Black Hole Spins." *Astrophysical Journal*, 620:69-77, 2005 February 10.

[153] King *et al.*, "The Evolution of Black Hole Mass and Spin in Active Galactic Nuclei." *Monthly Notices of the Royal Astronomical Society*, 385, 1621-1627 (2008).

[154] Maio *et al.*, "Effects of Circumnuclear Disk Gas Evolution on the Spin of Central Black Holes." *Astrophysical Journal*, 767:37-50, 2013 April 10.

[155] Volonteri *et al.*, "The Evolution of Active Galactic Nuclei and Their Spins." *Astrophysical Journal*, 775:94-107, 2013 October 1.

[156] Wang *et al.*, "Evidence for Rapidly Spinning Black Holes in Quasars." *Astrophysical Journal*, 642:L111-L114, 2006 April 10.

[157] Cao, X., "Growth of Massive Black Holes During Radiatively Inefficient Accretion Phases." *Astrophysical Journal*, 659:950-957, 2007 April 20.

[158] Tanaka *et al.*, "Gravitationally Redshifted Emission Implying an Accretion Disk and Massive Black Hole in the Active Galaxy MCG-6-30-15." *Nature*, 375:659-661, 22 June 1995.

[159] Brenneman, L.W. and Reynolds, C.S., "Constraining Black Hole Spin Via X-Ray Spectroscopy." *Astrophysical Journal*, 652:1028-1043, 2006 December 1.

[160] Jiang *et al.*, "Gemini Near-Infrared Spectroscopy of Luminous $z \sim 6$ Quasars: Chemical Abundances, Black Hole Masses, and Mg_{II} Absorption." *Astronomical Journal*, 134:1150-1161, 2007 September.

[161] Mashhoon, B., "On the Strength of a Gravitational Field." *Physics Letters A*, 163:7-14 (1992).

[162] Kopeikin, S. and Schafer, G., "Lorentz Covariant Theory of Light Propagation in Gravitational Fields of Arbitrary-Moving Bodies." *Physical Review D*, 60:124002, 5 November 1999.

[163] Klioner, S.A. and Peip, M., "Numerical Simulations of the Light Propagation in the Gravitational Field of Moving Bodies." *Astronomy & Astrophysics*, 410, 1063-1074 (2003).

[164] Kopeikin, S. and Mashhoon, B., "Gravitomagnetic Effects in the Propagation of Electromagnetic Waves in Variable Gravitational Fields of Arbitrary-Moving and Spinning Bodies." *Physical Review D*, 65:064025, 28 February 2002.

[165] Will, C., "Incorporating Post-Newtonian Effects in $\mathcal{N}$-Body Dynamics." *Physical Review D*, 89:044043, 26 February 2014.

[166] Nagar *et al.*, "Accretion-Driven Gravitational Radiation from Nonrotating Compact Objects: Infalling Quadrupolar Shells." *Physical Review D*, 69:124028, 28 June 2004.

[167] Zel'dovich, Y.B., "Electromagnetic and Gravitational Waves in a Stationary Magnetic Field." *Zh. Eksp. Teor. Fiz.* 65, 1311-1315, October 1973.

[168] Sigl *et al.*, "Gravitational-Wave Background from Compact Objects Embedded in Active Galactic Nuclei Accretion Disks." *Physical Review D*, 75:024034, 26 January 2007.

[169] Buonanno *et al.*, "Inspiral, Merger, and Ring-Down of Equal-Mass Black-Hole Binaries." *Physical Review D*, 75:124018, 21 June 2007.

[170] Baker *et al.*, "Consistency of Post-Newtonian Waveforms with Numerical Relativity." *Physical Review Letters*, 99:181101, 29 October 2007.

[171] Rubin *et al.*, "Rotational Properties of 21 Sc Galaxies with a Large Range of Luminosities and Radii, from NGC 4605 ($R = 4$ kpc) to UGC 2885 ($R = 122$ kpc)." *Astrophysical Journal*, 238:471-487, 1980 June 1.

[172] Battaglia *et al.*, "The Radial Velocity Dispersion Profile of the Galactic Halo: Constraining the Density Profile of the Dark Halo of the Milky Way." *Monthly Notices of the Royal Astronomical Society*, 364, 433-442 (2005).

[173] Webb, S., *New Eyes on the Universe* (Springer, 2012).

[174] Burke *et al.*, "Clusters Of Galaxies, X-Rays, and Dark Matter." in Spooner, N. and Kudryavtsev, V. (editors), *The Identification of Dark Matter*, (World Scientific Publishing Co., 1999).

[175] Romanowsky *et al.*, "A Dearth of Dark Matter in Ordinary Elliptical Galaxies." *Science*, Volume 301, Issue 5640, pp 1696-1698, 19 September 2003.

[176] Strigari, L., "Galactic Searches for Dark Matter." *Physics Reports*, Volume 531, Issue 1, pp 1-88 (2013).

[177] Clowe *et al.*, "A Direct Empirical Proof of the Existence of Dark Matter." *Astrophysical Journal*, 648:L109-L113, 2006 September 10.

[178] Markevitch *et al.*, "Direct Constraints on the Dark Matter Self-Interaction Cross Section from the Merging Galaxy Cluster 1E 0657-56." *Astrophysical Journal*, 606:819-824, 2004 May 10.

[179] Harvey *et al.*, "The Non-Gravitational Interactions of Dark Matter in Colliding Galaxy Clusters." *Science*, 347:6229, 27 March 2015.

[180] Harvey *et al.*, "Looking for Dark Matter Trails in Colliding Galaxy Clusters." *Monthly Notices of the Royal Astronomical Society*, 464:3991-3997, 1 February 2017.

[181] Mo, H., "The Link Between Galaxies and Dark Matter." *International Journal of Modern Physics D*, Volume 20, Issue 10, pp 1771-1777, September 2011.

[182] Navarro, J. and White, S., "Simulations of Dissipative Galaxy Formation in Hierarchically Clustering Universes - II. Dynamics of the Baryonic Component in Galactic Haloes." *Monthly Notices of the Royal Astronomical Society*, 267, 401-412, 15 March 1994.

[183] Klypin *et al.*, "Galaxies in $\mathcal{N}$-Body Simulations: Overcoming the Overmerging Problem." *Astrophysical Journal*, 516:530-551, 1999 May 10.

[184] Diemand *et al.*, "Formation and Evolution of Galaxy Dark Matter Halos and Their Substructure." *Astrophysical Journal*, 667:859-877, 2007 October 1.

[185] Gao *et al.*, "The Subhalo Populations of ΛCDM Dark Haloes." *Monthly Notices of the Royal Astronomical Society*, 355, 819-834 (2004).

[186] Springel *et al.*, "The Aquarius Project: The Subhaloes of Galactic Haloes." *Monthly Notices of the Royal Astronomical Society*, 391, 1685-1711 (2008).

[187] Mo, H. and Mao, S., "Galaxy Formation in Pre-Processed Dark Haloes." *Monthly Notices of the Royal Astronomical Society*, 353, 829-840 (2004).

[188] McIntosh *et al.*, "Ongoing Assembly of Massive Galaxies by Major Merging in Large Groups and Clusters from the SDSS." *Monthly Notices of the Royal Astronomical Society*, 388, 1537-1556 (2008).

[189] Hester, J.A. and Tasitsiomi, A., "Dark Matter Halo Mergers: Dependence on Environment." *Astrophysical Journal*, 715:342-354, 2010 May 20.

[190] Mihos, J. and Hernquist, L., "Dense Stellar Cores in Merger Remnants." *Astrophysical Journal*, 437:L47-50, 1994 December 10.

[191] Capelo, P. and Dotti, M., "Shocks and Angular Momentum Flips: A Different Path to Feeding the Nuclear Regions of Merging Galaxies." *Monthly Notices of the Royal Astronomical Society*, 465:3, 2643-2653 (2017).

[192] Hilz *et al.*, "Relaxation and Stripping - The Evolution of Sizes, Dispersions and Dark Matter Fractions in Major and Minor Mergers of Elliptical Galaxies." *Monthly Notices of the Royal Astronomical Society*, 425, 3119-3136 (2012).

[193] Mo, H., Van den Bosch, F. and White, S., *Galaxy Formation and Evolution* (Cambridge University Press, 2010).

[194] Stewart *et al.*, "Galaxy Mergers and Dark Matter Halo Mergers in ΛCDM: Mass, Redshift, and Mass-Ratio Dependence." *Astrophysical Journal*, 702:1005-1015, 2009 September 10.

[195] Stewart K., "Invisible Major Mergers: Why the Definition of a Galaxy 'Merger Ratio' Matters." *ASP Conference Series*, Vol. 419 (2010), S. Jogee, I. Marinova, L. Hao, and G.A. Blanc (editors).

[196] Mathewson, S., "Powerful Black Hole Found Within Unconventional Galaxy Merger." <http://www.space.com/36448-powerful-black-hole-unconventional-galaxy-merger.html>.

Chapter 5

[197] Villar-Martin *et al.*, "VLT and GTC Observations of SDSS J0123+00: A type 2 Quasar Triggered in a Galaxy Encounter?" *Monthly Notices of the Royal Astronomical Society Letters*, 407:L6-L10 (2010).

[198] Lopez-Corredoira, M. and Gutierrez, C.M., "Research on Candidates for Non-Cosmological Redshifts." *AIP Conference Proceedings*, 822:75 (2006).

[199] Burbidge, G., "Noncosmological Redshifts." *Publications of the Astronomical Society of the Pacific*, 113, 899-902 (2001).

[200] Peterson, B.M., "Measuring the Masses of Supermassive Black Holes." *Space Science Reviews*, Volume 183, Issue 1-4, pp 253-275, September 2014.

[201] McLure, R. and Dunlop, J., "The Cosmological Evolution of Quasar Black Hole Masses." *Monthly Notices of the Royal Astronomical Society*, 352, 1390-1404 (2004).

[202] Corbett *et al.*, "Emission Linewidths and QSO Black Hole Mass Estimates from the 2dF QSO Redshift Survey." *Monthly Notices of the Royal Astronomical Society*, 343, 705-718 (2003).

[203] Mejia-Restrepo *et al.*, "Active Galactic Nuclei at $z \sim 1.5$ - II. Black Hole Mass Estimation by Means of Broad Emission Lines." *Monthly Notices of the Royal Astronomical Society*, 460, 187-211 (2016).

[204] Trakhtenbrot, B. and Netzer, H., "Black Hole Growth to $z = 2$ - I. Improved Virial Methods for Measuring MBH and L/L_{Edd}." *Monthly Notices of the Royal Astronomical Society*, 427, 3081-3102 (2012).

[205] Netzer, H. and Trakhtenbrot, B., "Cosmic Evolution of Mass Accretion Rate and Metallicity in Active Galactic Nuclei." *Astrophysical Journal*, 654:754-763, 2007 January 10.

[206] Shen, Y. and Kelly, B., "The Impact of the Uncertainty in Single-Epoch Virial Black Hole Mass Estimates on the Observed Evolution of the Black Hole-Bulge Scaling Relations." *Astrophysical Journal*, 713:41-45, 2010 April 10.

[207] Trakhtenbrot *et al.*, "Faint COSMOS AGNs at $z \sim 3.3$. I. Black Hole Properties and Constraints on Early Black Hole Growth." *Astrophysical Journal*, 825:4-20, 2016 July 1.

[208] Marchesi *et al.*, "The Chandra COSMOS-Legacy Survey: The $z > 3$ Sample." *Astrophysical Journal*, 827:150-165, 2016 August 20.

[209] Netzer, H., The Largest Black Holes and the Most Luminous Galaxies." *Astrophysical Journal*, 583:L5-L8, 2003 January 20.

[210] Frank J., King, A.R. and Raine, D., *Accretion Power in Astrophysics* (Cambridge University Press, 2002).

[211] Netzer, H., *The Physics and Evolution of Active Galactic Nuclei* (Cambridge University Press, 2013), Chapter 3.

[212] Fine *et al.*, "The Evolution of Host Mass and Black Hole Mass in Quasi-Stellar Objects from the 2dF QSO Redshift Survey." *Monthly Notices of the Royal Astronomical Society*, 373, 613-622 (2006).

[213] Shen, Y. and Kelly, B., "The Demographics of Broad-Line Quasars in the Mass-Luminosity Plane. I. Testing FWHM-Based Virial Black Hole Masses." *Astrophysical Journal*, 746:169-190, 2012 February 3.

[214] Kelly, B. and Shen, Y., "The Demographics of Broad-Line Quasars in the Mass-Luminosity Plane. II. Black Hole Mass and Eddington Ratio Functions." *Astrophysical Journal*, 764:45-69, 2013 February 10.

[215] Shen, Y., "The Mass of Quasars." *Bulletin of the Astronomical Society of India*, 41:61-115 (2013).

[216] Schulze *et al.*, "The Cosmic Growth of the Active Black Hole Population at $1 < z < 2$ in zCOSMOS, VVDS and SDSS." *Monthly Notices of the Royal Astronomical Society*, 447, 2085-2111 (2015).

[217] Laor, A. and Netzer, H., "Massive Thin Accretion Discs - I. Calculated Spectra." Monthly *Notices of the Royal Astronomical Society*, 238, 897-916 (1989).

[218] Shlosman *et al.*, "The Fuelling of Active Galactic Nuclei." *Nature* 345:679-686, 21 June 1990.

[219] Abramowicz, M. and Fragile, P., "Foundations of Black Hole Accretion Disk Theory." *Living Reviews in Relativity*, 16:1, December 2013.

[220] Falanga, Belloni, Casella, Gilfanov, Jonker and King (editors), *The Physics of Accretion onto Black Holes* (Springer, 2015).

[221] Narayan, R. and Quataert, E., "Black Hole Accretion." *Science*, Volume 307, Issue 5706, pp 77-80, 7 January 2005.

[222] Begelman, M.C. and Meier, D.L., "Thick Accretion Disks: Self-similar, Supercritical Models." *Astrophysical Journal*, 253:873-896, 1982 February 15.

[223] Collin *et al.*, "Are Quasars Accreting at Super-Eddington Rates?" *Astronomy & Astrophysics*, 388, 771-786 (2002).

[224] Abramowicz *et al.*, "Slim Accretion Disks." *Astrophysical Journal*, 332:646-658, 1988 September 15.

[225] Collin, S. and Kawaguchi, T., "Super-Eddington Accretion Rates in Narrow Line Seyfert 1 Galaxies." *Astronomy & Astrophysics*, 426, 797-808 (2004).

[226] Narayan, R. and Yi, I., "Advection-Dominated Accretion: A Self-Similar Solution." *Astrophysical Journal*, 428:L13-L16, 1994 June.

[227] Narayan *et al.*, "Global Structure and Dynamics of Advection-Dominated Accretion Flows Around Black Holes." *Astrophysical Journal*, 476:49-60, 1997 February 10.

[228] Narayan *et al.*, "Advection-Dominated Accretion and Black Hole Event Horizons." *Astrophysical Journal*, 478:L79-82, 1997 April 1.

[229] Narayan *et al.*, "Self-Similar Accretion Flows with Convection." *Astrophysical Journal*, 539:798-808, 2000 August 20.

[230] Blandford, R. and Begelman, M. "On the Fate of Gas Accreting at a Low Rate on to a Black Hole." *Monthly Notices of the Royal Astronomical Society*, 303, L1-L5 (1999).

[231] Quataert, E. and Gruzinov, A., "Convection-Dominated Accretion Flows." *Astrophysical Journal*, 539:809-814, 2000 August 20.

[232] Netzer, H. and Trakhtenbrot, B., "Bolometric Luminosity Black Hole Growth Time and Slim Accretion Discs in Active Galactic Nuclei." *Monthly Notices of the Royal Astronomical Society*, 438, 672-679 (2014).

[233] Trakhtenbrot, B., "The Most Massive Active Black Holes at z ~ 1.5-3.5 have High Spins and Radiative Efficiencies." *Astrophysical Journal*, 789:L9-L13, 2014 July 1.

[234] King, A., "Black Holes, Galaxy Formation, and the M_{BH}-σ Relation." *Astrophysical Journal*, 596:L27-L29, 2003 October 20.

[235] Ricci *et al.*, "Growing Supermassive Black Holes in the Late Stages of Galaxy Mergers are Heavily Obscured." *Monthly Notices of the Royal Astronomical Society*, 468, 1273-1299 (2017).

[236] Munch, G., Mampaso, A. and Sanchez, F. (editors), *The Universe at Large: Key Issues in Astronomy and Cosmology* (Cambridge University Press, 1997), page 296.

[237] Krolik, J.H., "Magnetized Accretion Inside the Marginally Stable Orbit Around a Black Hole." *Astrophysical Journal*, 515:L73-L76, 1999 April 20.

[238] Zakharov *et al.*, "Magnetic Fields in Active Galactic Nuclei and Microquasars." *Monthly Notices of the Royal Astronomical Society*, 342, 1325-1333 (2003).

[239] Zhang *et al.*, "The Black Hole Mass and Magnetic Field Correlation in Active Galactic Nuclei." *Chinese Journal of Astronomy and Astrophysics*, Volume 5 Supp., 347 (2005).

[240] Silant'ev, N.A. *et al.*, "Magnetic Fields of Active Galactic Nuclei and Quasars with Regions of Polarized Broad Hα Lines." *Astrophysical Bulletin*, 68:14-25, January 2013.

[241] Mineshige *et al.*, "Gravitational-Wave Radiation from Magnetized Accretion Disks." *Publications of the Astronomical Society of Japan*, Volume 54, 655-660, 2002 October 25.

[242] Lai, D., "Physics in Very Strong Magnetic Fields." *Space Science Reviews*, Volume 191, Issue 1-4, pp 13-25, October 2015.

[243] Potekhin, A., "Structure and Radiative Transitions of the Hydrogen Atom Moving in a Strong Magnetic Field." *Journal of Physics B: Atomic, Molecular and Optical Physics*, 27:1073 (1994).

[244] Lai, D., "Matter in Strong Magnetic Fields." *Reviews of Modern Physics*, 73:629, July 2001.

[245] Skjervold, J. and Ostgaard, E., "Heavy Atoms in Superstrong Magnetic Fields." *Physica Scripta*, Vol. 29, 543-550 (1984).

[246] Robson, I., *Active Galactic Nuclei* (John Wiley & Sons, 1996), pp 206-207.

[247] Tytler, D. and Fang, X., "Systematic QSO Emission-Line Velocity Shifts and New Unbiased Redshifts." *Astrophysical Journal Supplement Series*, 79:1-36, 1992 March.

[248] Sulentic *et al.*, "Eigenvector 1: An Optimal Correlation Space for Active Galactic Nuclei." *Astrophysical Journal*, 536:L5-L9, 2000 June 10.

[249] Gaskell, C., "What Broad Emission Lines tell us about how Active Galactic Nuclei Work." *New Astronomy Reviews*, 53:140-148 (2009).

[250] Gaskell, C. and Goosmann, R., "The Case for Inflow of the Broad-Line Region of Active Galactic Nuclei." *Astrophysics & Space Science*, 361:67, February 2016.

[251] Beckmann, V. and Shrader, C., *Active Galactic Nuclei* (Wiley-VCH, 2012), pp 272-276.

[252] Djorgovski *et al.*, "On the Threshold of the Reionization Epoch." *Astrophysical Journal*, 560:L5-L8, 2001 October 10.

[253] Becker *et al.*, "Evidence for Reionization at z ~ 6: Detection of a Gunn-Peterson Trough in a z = 6.28 Quasar." *Astronomical Journal*, 122:2850-2857, 2001 December.

[254] Oh, S.P. and Furlanetto, S.R., "How Universal is the Gunn-Peterson Trough at z ~ 6?: A Closer Look at the Quasar SDSS J1148+5251." *Astrophysical Journal*, 620:L9-L12, 2005 February 10.

[255] Fan *et al.*, "Constraining the Evolution of the Ionizing Background and the Epoch of Reionization with z ~ 6 Quasars II: A Sample of 19 Quasars." *Astronomical Journal*, 132:117-136, 2006 July.

[256] Tauber *et al.*, "Planck 2013 Results. I. Overview of Products and Scientific Results." *Astronomy & Astrophysics*, 571:A1 (2014).

[257] Ciardi, B. and Ferrara, A., "The First Cosmic Structures and Their Effects." *Space Science Reviews*, Volume 116, Issue 3-4, pp 625-705, February 2005.

[258] Jakobsen *et al.*, "Detection of Intergalactic Ionized Helium Absorption in a High-Redshift Quasar." *Nature*, 370:35-39, 7 July 1994.

[259] Syphers *et al.*, "Ten More New Sightlines for the Study of Intergalactic Helium, and Hundreds of Far-UV-Bright Quasars, from SDSS, *GALEX* and *HST*." *Astrophysical Journal Supplement Series*, 185:20-31, 2009 September.

[260] Syphers *et al.*, "*HST*/COS Observations of Thirteen New He_{II} Quasars." *Astronomical Journal*, 143:100-111, 2012 April.

[261] Shull *et al.*, "*HST*/COS Observations of the Quasar HE 2347-4342: Probing the Epoch of He_{II} Patchy Reionization at Redshifts $z = 2.4$-2.9." *Astrophysical Journal*, 722:1312-1324, 2010 October 20.

[262] Worseck *et al.*, "Early and Extended Helium Reionization over more than 600 Million Years of Cosmic Time." *Astrophysical Journal*, 825:144-175, 2016 July 10.

[263] Worseck *et al.*, "The End of Helium Reionization at $z \sim 2.7$ Inferred from Cosmic Variance in *HST*/COS He_{II} Lyα Absorption Spectra." *Astrophysical Journal*, 733:L24-29, 2011 June 1.

[264] Hao *et al.*, "Cosmic Reionization of Hydrogen and Helium: Contribution from both Mini-Quasars and Stars." *Monthly Notices of the Royal Astronomical Society*, 451, 1875-1882 (2015).

Chapter 6

[265] Corbin, M.R., "On the Relation of the X-Ray and Emission-Line Properties of Low-Redshift QSOs." *Astrophysical Journal*, 403:L9-L11, 1993 January 20.

[266] Wang, T., Brinkmann, W. and Bergeron, J., "X-Ray Properties of Active Galactic Nuclei with Optical Fe_{II} Emission." *Astronomy & Astrophysics*, 309, 81-96 (1996).

[267] Corbin, M.R. and Boroson, T.A., "Combined Ultraviolet and Optical Spectra of 48 Low-Redshift QSOs and the Relation of the Continuum and Emission-Line Properties." *Astrophysical Journal Supplement Series*, 107:69-96, 1996 November.

[268] Corbin, M.R., "Relativistic Effects in the QSO Broad-Line Region." *Astrophysical Journal*, 485:517-522, 1997 August 20.

[269] Kormendy *et al.*, "Structure and Formation of Elliptical and Spheroidal Galaxies." *Astrophysical Journal Supplement Series*, 182:216-309, 2009 May.

[270] McLure, R. and Jarvis, M., "The Relationship Between Radio Luminosity and Black-Hole Mass in Optically-Selected Quasars." *Monthly Notices of the Royal Astronomical Society*, 353, L45-L49, 1 October 2004.

[271] Sikora, *et al.*, "Radio Loudness of Active Galactic Nuclei: Observational Facts and Theoretical Implications." *Astrophysical Journal*, 658:815-828, 2007 April 1.

[272] Heckman *et al.*, "Galaxy Collisions and Mergers: The Genesis of Very Powerful Radio Sources?" *Astrophysical Journal*, 311:526-547, 1986 December 15.

[273] Becker *et al.*, "Properties of Radio-Selected Broad Absorption Line Quasars from the First Bright Quasar Survey." *Astrophysical Journal*, 538:72-82, 2000 July 20.

[274] Gregg *et al.*, "FR II Broad Absorption Line Quasars and the Life Cycle of Quasars." *Astrophysical Journal*, 641:210-216, 2006 April 10.

[275] Trump *et al.*, "Accretion Rate and the Physical Nature of Unobscured Active Galaxies." *Astrophysical Journal*, 733:60-73, 2011 May 20.

[276] Plotkin *et al.*, "Using the Fundamental Plane of Black Hole Activity to Distinguish X-ray Processes from Weakly Accreting Black Holes." *Monthly Notices of the Royal Astronomical Society*, 419:267-286, 1 January 2012..

[277] Merloni *et al.*, "Why the Fundamental Plane of Black Hole Activity is not Simply a Distance Driven Artifact." *New Astronomy*, Volume 11, Issue 8, pp 567-576, July 2006.

[278] Marziani *et al.*, "Searching for the Physical Drivers of the Eigenvector 1 Correlation Space." *Astrophysical Journal*, 558:553-560, 2001 September 10.

[279] Sulentic *et al.*, "Phenomenology of Broad Emission Lines in Active Galactic Nuclei." *Annual Review of Astronomy & Astrophysics*, 38:521-571 (2000).

[280] Marziani *et al*, "An Optical Spectroscopic Atlas of Low-Redshift Active Galactic Nuclei." *Astrophysical Journal Supplement Series*, 145:199-211, 2003 April.

[281] Sulentic *et al.*, "Average Quasar Spectra in the Context of Eigenvector 1." *Astrophysical Journal*, 566:L71-L75, 2002 February 20.

[282] Zamanov *et al.*, "Kinematic Linkage Between the Broad- and Narrow-Line-Emitting Gas in Active Galactic Nuclei." *Astrophysical Journal*, 576:L9-L13, 2002 September 1.

[283] Sulentic *et al.*, "C_{IV} λ1549 as an Eigenvector 1 Parameter for Active Galactic Nuclei." *Astrophysical Journal*, 666:757-777, 2007 September 10.

[284] Laor, A., "On Black Hole Masses and Radio Loudness in Active Galactic Nuclei." *Astrophysical Journal*, 543:L111-L114, 2000 November 10.

[285] Sulentic *et al.*, "The Case for Two Quasar Populations." *Open Astronomy*, 20:427-434 (2011).

[286] Boroson, T., "Black Hole Mass and Eddington Ratio as Drivers for the Observable Properties of Radio-Loud and Radio-Quiet QSOs." *Astrophysical Journal*, 565:78-85, 2002 January 20.

[287] Marziani *et al.*, "Searching for the Physical Drivers of Eigenvector 1: Influence of Black Hole Mass and Eddington Ratio." *Monthly Notices of the Royal Astronomical Society*, 345, 1133-1144, 11 November 2003.

[288] Zamfir *et al.*, "New Insights on the QSO Radio-Loud/Radio-Quiet Dichotomy: SDSS Spectra in the Context of the 4D Eigenvector1 Parameter Space." *Monthly Notices of the Royal Astronomical Society*, 387, 856-870, 21 June 2008.

[289] Krongold *et al.*, "The Circumgalactic Environment of Bright *IRAS* Galaxies." *Astrophysical Journal*, 572:169-177, 2002 June 10.

[290] Villarroel, B. and Korn, A.J., "The Different Neighbours Around Type-1 and Type-2 Active Galactic Nuclei." *Nature Physics*, 10:417-420, 11 May 2014.

[291] Zamfir *et al.*, "Detailed Characterization of Hβ Emission Line Profile in Low-z SDSS Quasars." *Monthly Notices of the Royal Astronomical Society*, 403, 1759-1786 (2010).

[292] Marziani *et al.* "UV Spectral Diagnostics for Low Redshift Quasars: Estimating Physical Conditions and Radius of the Broad Line Region." *Astrophysics & Space Science*, Volume 356, Issue 2, pp 339-346, April 2015.

[293] Bensch *et al.*, "Measures of the Soft X-ray Excess as an Eigenvector 1 Parameter for Active Galactic Nuclei." *Journal of Astrophysics and Astronomy*, Volume 36, Issue 4, pp 467-474, December 2015.

[294] Marziani, P. and Sulentic, J., "Quasars and Their Emission Lines as Cosmological Probes." *Advances in Space Research*, 54:7, 1331-1340, 1 October 2014.

[295] Fraix-Burnet *et al.*, "The Phylogeny of Quasars and the Ontogeny of Their Central Black Holes." *Frontiers in Astronomy and Space Sciences*, Volume 4, 27 February 2017.

[296] Dultzin *et al.*, "Narrow-Line Seyfert 1s: A Luminosity Dependent Definition." *Proceedings of Science*, April 2011.

[297] Negrete *et al.*, "Broad-Line Region Physical Conditions in Extreme Population A Quasars: A Method to Estimate Central Black Hole Mass at High Redshift." *Astrophysical Journal*, 757:62-76, 2012 September 20.

[298] Sulentic *et al.*, "VLT/ISAAC Spectra of the Hβ Region in Intermediate Redshift Quasars." *Astronomy & Astrophysics*, 423, 121-132 (2004).

[299] Sun, J. and Shen, Y., "Dissecting the Quasar Main Sequence: Insight from Host Galaxy Properties." *Astrophysical Journal*, 804:L15-L19, 2015 May 1.

[300] Marziani *et al.*, "VLT/ISAAC Spectra of the Hβ Region in Intermediate-Redshift Quasars III. Hβ Broad-Line Profile Analysis and Inferences about BLR Structure." Astronomy & Astrophysics, 495, 83-112 (2009).

[301] Marziani *et al.* "Blue Outliers Among Intermediate Redshift Quasars." *Astrophysics & Space Science*, 361:3, January 2016.

[302] Shen, Y. and Ho, L., "The Diversity of Quasars Unified by Accretion and Orientation." *Nature*, 513:210-213, 11 September 2014.

[303] Bachev *et al.*, "Average Ultraviolet Quasar Spectra in the Context of Eigenvector 1: A Baldwin Effect Governed by the Eddington Ratio?" *Astrophysical Journal*, 617:171-183, 2004 December 10.

[304] Marziani, P. and Sulentic, J., "Quasar Outflows: in the 4D Eigenvector 1 Context." *Astronomical Review*, 7:4, pp 33-57 (2012).

[305] Marziani *et al.* "The Most Powerful Quasar Outflows as Revealed by the $C_{IV}\lambda1549$ Resonance Line." *Astrophysics & Space Science*, 361:29, January 2016.

[306] Sulentic *et al.*, "GTC Spectra of $z \approx 2.3$ Quasars: Comparison with Local Luminosity Analogs." *Astronomy & Astrophysics*, 570:A96 (2014).

[307] Sulentic *et al.*, "What does $C_{IV}\lambda1549$ tell us About the Physical Driver of the Eigenvector Quasar Sequence?" Astro*nomy & Astrophysics*, 608:A122, December 2017.

[308] Risaliti *et al.*, "The Sloan Digital Sky Survey/*XMM-Newton* Quasar Survey: Correlation Between X-Ray Spectral Slope and Eddington Ratio." *Astrophysical Journal*, 700:L6-L10, 2009 July 20.

[309] Brightman *et al.*, "A Statistical Relation Between the X-Ray Spectral Index and Eddington Ratio of Active Galactic Nuclei in Deep Surveys." *Monthly Notices of the Royal Astronomical Society*, 433:3 pp 2485-2496, 11 August 2013.

[310] Marziani *et al.*, "Low- and High-z Highly Accreting Quasars in the 4D Eigenvector 1 Context." *Astronomical Review*, 9:4, pp 1-20 (2014).

[311] Brightman *et al.*, "Compton Thick Active Galactic Nuclei in *Chandra* Surveys." *Monthly Notices of the Royal Astronomical Society*, 443, pp 1999-2017 (2014).

[312] Trakhtenbrot *et al.*, "ALMA Observations Show Major Mergers Among the Host Galaxies of Fast-Growing High-Redshift, Supermassive Black Holes." *Astrophysical Journal*, 836:1, pp 8-28, 2017 February 10.

[313] Kocevski *et al.*, "Are Compton-Thick AGNs the Missing Link Between Mergers and Black Hole Growth?" *Astrophysical Journal*, 814:2, pp 104-115, 2015 December 1.

[314] Schneider *et al.*, "The Sloan Digital Sky Survey Quasar Catalog. III. Third Data Release." *Astronomical Journal*, 130:367-380, 2005 August.

[315] Jiang *et al.*, "The Radio-Loud Fraction of Quasars is a Strong Function of Redshift and Optical Luminosity." *Astrophysical Journal*, 656:680-690, 2007 February 20.

Appendix A

[316] Cheng, Ta-Pei, *Relativity, Gravitation and Cosmology* (Oxford University Press, 2010).

[317] Napier, W.M., and Guthrie, B.N.G., "Quantized Redshifts: A Status Report" *Astronomy & Astrophysics*, 18:455-463 (1997).

Appendix B

[318] Sulentic, J., "Toward a Classification Scheme for Broad-Line Profiles in Active Galactic Nuclei." *Astrophysical Journal*, 343:54-65, 1989 August 1.

[319] Marziani *et al.*, "The Peculiar Balmer Line Profiles of OQ 208." *Astrophysical Journal*, 410:56-83, 1993 January 7.

[320] Marziani *et al.*, "Comparative Analysis of the High- and Low-Ionization Lines in the Broad-Line Region of Active Galactic Nuclei." *Astrophysical Journal Supplement Series*, 104:37-70, 1996 May.

[321] Sulentic *et al.*, "On the Origin of Broad Fe Kα and H_I Hα Lines in Active Galactic Nuclei." *Astrophysical Journal*, 501:54-68, 1998 July 1.

[322] Sulentic, J. and Marziani, P., "The Intermediate-Line Region in Active Galactic Nuclei: A Region "Paeter Necessitatem." *Astrophysical Journal*, 518:L9-L12, 1999 June 10.

[323] Sulentic *et al.*, "The Demise of the Classical Broad-Line Region in the Luminous Quasar PG 1416-129." *Astrophysical Journal*, 545, pp L15-L18, 2000 December 10.

[324] Zamanov R, and Marziani, P., "Searching for the Physical Drivers of Eigenvector 1: From Quasars to Nanoquasars." *Astrophysical Journal*, 571:L77-L80, 2002 June 1.

[325] Sulentic *et al.*, "Radio-Loud Active Galactic Nuclei in the Context of the Eigenvector 1 Parameter Space." *Astrophysical Journal*, 597:L17-L20, 2003 November 1.

[326] Sulentic *et al.*, "Our Search for an H-R diagram of Quasars." *Revisita Mexicana de Astronomia y Astrofisica Conference Series*, Vol. 32, pp 51-58 (2008).

[327] Sulentic *et al.*, "Comparing Hβ Line Profiles in the 4D Eigenvector 1 Context." *New Astronomy Reviews*, 53:198-201 (2009).

[328] Marziani *et al.*, "Broad-Line Region Physical Conditions Along the Quasar Eigenvector 1 Sequence." *Monthly Notices of the Royal Astronomical Society*, 409:3, pp 1033-1048, 11 December 2010.

[329] Sulentic *et al.*, "No Evidence for a Systematic Fe II Emission Line Redshift in Type 1 Active Galactic Nuclei." *Astrophysical Journal*, 752:L7-L11, 2012 June 10.

[330] Marziani *et al.*, "Low-Ionization Outflows in High Eddington Ratio Quasars." *Astrophysical Journal*, 764:150-159, 2013 February 20.

[331] Marziani, P. and Sulentic, J., "Highly Accreting Quasars: Sample Definition and Possible Cosmological Implications." *Monthly Notices of the Royal Astronomical Society*, 442:2, pp 1211-1229, 1 August 2014.

[332] Sulentic *et al.*, "Fifty Years of Quasars: Physical Insights and Potential for Cosmology." *Journal of Physics: Conference Series,* Volume 565, 25 Nov 2014.

[333] Sulentic, J. and Marziani, P., "Quasars in the 4D Eigenvector 1 Context: A Stroll Down Memory Lane." *Frontiers in Astronomy and Space Sciences,* Volume 2, October 2015.

[334] Marziani *et al.* "A Main Sequence for Quasars." *Frontiers in Astronomy and Space Sciences,* Volume 1, March 2018.

Made in United States
Troutdale, OR
11/03/2026

56844433R00114